Lecture Notes in Mathematics

Edited by A. Dold and B. Eckmann

Subseries: Tata Institute of Fundamental Research, Bombay
Adviser: M.S. Narasimhan

156

Robin Hartshorne

Ample Subvarieties of Algebraic Varieties

Notes written in Collaboration with C. Musili

Springer-Verlag
Berlin Heidelberg New York Tokyo

Author

Robin Hartshorne
Department of Mathematics, University of California, Berkeley
Berkeley, CA 94720, USA

1st Edition 1970
2nd Printing 1986

ISBN 3-540-05184-8 Springer-Verlag Berlin Heidelberg New York Tokyo
ISBN 0-387-05184-8 Springer-Verlag New York Heidelberg Berlin Tokyo

Printing and binding: Beltz Offsetdruck, Hemsbach/Bergstr.
2146/3140-543210

PREFACE

These notes are an enlarged version of a three-month course of lectures I gave at the Tata Institute of Fundamental Research during the winter of 1969-70, while on sabbatical leave from Harvard. Their style is informal. I hope they will serve as an introduction to some current research topics, for students who have had a one year course in modern algebraic geometry.

They contain some known material, mostly available only in original research papers, and some new material published here for the first time. In particular, I have added Ch. I §10, Ch. II §5, Ch. III §§6,7,8, and Ch. VI, which contain new results mentioned only briefly in the lectures. For the benefit of future researchers there are a number of exercises, open problems, and conjectures. There is an extensive bibliography at the end.

I wish to thank the Tata Institute for inviting me to come to Bombay, and the auditors of the lectures for their questions and stimulating conversation. My special thanks go to C. Musili who recorded the lectures, wrote them up for inclusion in these notes, and assisted in all aspects of the preparation of the manuscript.

Finally, I wish to thank my wife, whose support has been invaluable.

Robin Hartshorne
Bombay
April, 1970

CONTENTS

INTRODUCTION

My intention in giving this course was to develop a notion of ample subvariety of an algebraic variety, which should generalize to higher codimensions the concept of an ample divisor. So far no single property of a subvariety has appeared to give an appropriate definition of "ampleness". So instead we have considered a number of different properties, each one of which generalizes in one way or another the notion of an ample divisor, and we have attempted to clarify the relations between these properties.

The first part of the course, consisting of Chapters I and II, is a review of the codimension one case. For divisors, we have a very beautiful situation where so many different approaches all lead to the same notion of an ample divisor. The geometric condition says that some multiple of the divisor should move in a sufficiently large linear system to give a projective embedding of the ambient variety. The cohomological condition says that tensoring with high multiples of the associated invertible sheaf should kill cohomology groups. The numerical condition (Nakai's criterion) says that the divisor and its self-intersections should be numerically positive.

In Chapter I we develop the theory of ample divisors, and give a proof of Nakai's criterion, following Kleiman. We also give a new criterion for an ample divisor, due to Seshadri, and we include Kleiman's solution of Chevalley's conjecture. In an appendix we give a new proof of an example of Mumford, of a numerically positive divisor which is not ample. Also we give an example of Ramanujam, based on the example of Mumford, of an effective numerically positive divisor which is not ample. The main sources for this chapter are Grothendieck [EGA, II, III] and Kleiman [2]. For more details, see the introduction to Chapter I.

In Chapter II we study the question of when the complement of a divisor is affine, following Goodman [1]. One sees easily that the complement of an ample divisor is affine. However the converse is not true. So we attempt to give weaker conditions on the divisor which should be equivalent to its complement being affine. We find that if $|D|$ is base-point free, then U is affine, where $U = X - \text{Supp } D$. Conversely, if U is an open affine subset of a complete non-singular variety X, then $Y = X - U$ is connected and has pure codimension one. In §5 we give a new result, which says that if $|nD|$ does not have too many base points (in a certain sense) for n large, then $X - \text{Supp } D$ is affine. We do not have any necessary and sufficient condition on D for $X - \text{Supp } D$ to be affine. However, there is the theorem of Goodman, proved in §6, which says that $U = X - Y$ is affine if and only if there is a birational blowing-up $f: \overline{X} \longrightarrow X$, with center contained in Y, such that $\overline{Y} = f^{-1}(Y)$ is the support of an ample divisor on $\overline{X}$.

The second part of the course, consisting of Chapters III, IV, and V, deals with subvarieties of higher codimension. The strongest condition we can put on a subvariety is to require that it be a complete intersection. A weaker, and purely local condition is to require that the normal bundle to the subvariety be ample, in the sense of ample vector bundles (see Chapter III, §1). Both of these conditions imply certain theorems on the vanishing or finite-dimensionality of the cohomology of the open complement of the subvariety. We define the cohomological dimension, cd(U), for a scheme U, to be the largest integer n such that $H^n(U,F) \neq 0$ for some coherent sheaf F on U. We define $q(U)$ to be the largest integer n (or -1) for which $H^n(U,F)$ is infinite-dimensional for some coherent sheaf F. A large part of Chapter III is devoted to the study of these integers. In case the ambient space is the projective space itself, Y is a closed subset, and U is the open complement, we have a fairly complete determination of $cd(U)$ and $q(U)$ in terms of various cohomological properties of Y. See §5 for statements of these results. The main sources for the early part of Chapter III are Hartshorne [AVB] and [CDAV]. The results of §§5,6,7,8 are mostly new, and have not been previously published.

There are a number of classical theorems of Lefschetz relating the fundamental group, Picard group, and complex cohomology of a projective variety to those of its hyperplane section. These can be generalized to ample divisors and to complete intersections of higher codimension. In Chapter IV we give a new simplified proof of Grothendieck's form of these theorems for π_1 and Pic. They can be thought of as giving another condition for "ampleness" of a subvariety. This chapter is based on Grothendieck's seminar [SGA 2].

Hironaka and Matsumura have found that the field of formal-rational functions along an ample divisor on a variety X, or along any subvariety of projective space, is isomorphic to the function field of the ambient variety itself. This suggests the properties

G2 and G3 of a subvariety: Y is G2 in X if the field $K(\hat{X})$ of formal-rational functions along Y is a finite algebraic extension of $K(X)$; Y is G3 in X if $K(\hat{X}) = K(X)$. We study these properties in Chapter V and prove their theorems there. The main sources for this chapter are Hartshorne [CDAV], Hironaka and Matsumura [1], and Speiser [1].

The main implications between all these different notions of "ampleness" for a subvariety are summarized in the table at the end of this introduction.

One important property of ample divisors which we have not attempted to generalize, and which is an interesting topic for future investigation, is the geometric one. If a subvariety moves in a "sufficiently large" algebraic family, it should have some of the other ampleness properties, and conversely. Closely connected to this is the question, to what extent are the properties we have considered stable under rational equivalence of subvarieties?

The third part of the course is Chapter VI, where we give a brief survey of analogous developments in the theory of several complex variables, and we prove two new comparison theorems for algebraic and analytic cohomology, generalizing Serre's [GAGA]. (Due to lack of time, these results were not presented in the oral lectures.)

MAIN IMPLICATIONS OF CHAPTERS III, IV, V.

Notations. X a non-singular projective variety of dimension n, Y a non-singular closed subvariety of dimension $s \geq 1$, $U = X - Y$.

Note also that for the case $X = \mathbb{P}^n_k$, the conditions $N_{Y/X}$ ample, G3, and $cd(U) < n-1$ always hold.

The symbol A, B ⟹ C means A+B ⟹ C. With arrows on both ends, it means A+B ⟺ C.

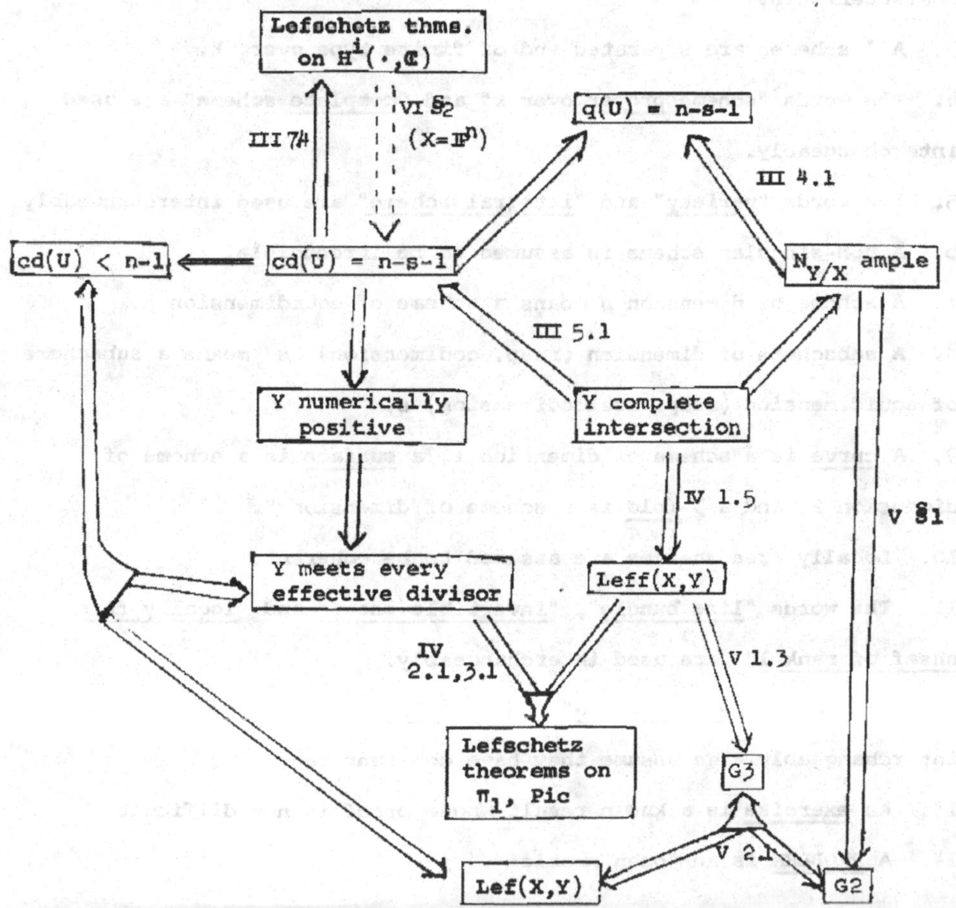

N.B. This diagram should be taken with a few grains of salt. For complete statements of results, see the text.

Notations, Terminology and Conventions.

Except at a few places, we have followed the notations and terminology of the main sources: notably Grothendieck [EGA] and [SGA].

Unless otherwise explicitly stated to the contrary, the following conventions will be in force throughout these notes.

1. All rings are assumed to be commutative with unit and noetherian.
2. k will denote an algebraically closed field of arbitrary characteristic.
3. All schemes are separated and of finite type over k.
4. The words "scheme proper over k" and "complete scheme" are used interchangeably.
5. The words "variety" and "integral scheme" are used interchangeably.
6. A non-singular scheme is assumed to be irreducible.
7. A scheme of dimension n means a scheme of equidimension n.
8. A subscheme of dimension (resp. codimension) s means a subscheme of equidimension (resp. equicodimension) s.
9. A curve is a scheme of dimension 1, a surface is a scheme of dimension 2, and a 3-fold is a scheme of dimension 3.
10. Locally free sheaves are assumed to be coherent.
11. The words "line bundle", "invertible sheaf" and "locally free sheaf of rank 1" are used interchangeably.
12. The words "vector bundle" and "locally free sheaves" are used interchangeably. We assume they have constant rank.
13. An exercise is a known result whose proof is not difficult.
14. A PROBLEM is an "open problem".

15. A CONJECTURE is a problem together with an opinion as to its probable outcome.

16. References and Cross-references:

a) When we want to refer to Theorem m.n of Chapter N, we will say "by Theorem m.n, Chapter N" or simply "by Theorem m.n" if the reference is made within Chapter N.

b) Numbers (resp. abbreviations) in the square brackets [ ] refer to the serial number (resp. abbreviations) in the Bibliography under the respective author.

Table of Conjectures and Open Problems

Chapter	Section	Conjectures	Open Problems
I	§10	....	10.7
II	§5	5.2	....
III	§2	....	2.3
	§3	3.10	3.14
	§4	4.4, 4.5, 4.16	4.6
	§5	5.4, 5.6, 5.7, 5.17	5.3, 5.5, 5.19
IV	§1	....	1.9
	§4	....	4.1
V	§4	....	4.7, 4.8
VI	§1	....	1.1
	§2	2.5	2.6
	§3	....	3.4

CHAPTER I

AMPLE DIVISORS

In this chapter we develop the theory of ample divisors. Roughly speaking, a divisor is ample if it moves in a sufficiently large linear system. The prototype of an ample divisor is the hyperplane section of a variety embedded in projective space. As the hyperplane moves, the divisor it cuts out moves in a linear system which separates points and also separates infinitely near points. Actually, the notion of ample divisor is somewhat broader than this, so we say a divisor D on X is _very ample_ if it is the divisor of hyperplane sections of X for some projective embedding of X. We say D is _ample_ if nD is very ample for some $n > 0$.

Thus our definition of ampleness is a geometric one. In the first few sections of this chapter we recall basic definitions and results about divisors, linear systems, and the associated rational map into projective space. In §2 we give a characterization of very ample divisors: D is very ample if and only if its complete linear system $|D|$ has no base points, separates points, and separates infinitely near points.

In §3 we use sheaf theory to give two characterizations of ample divisors. One is a sheaf-theoretic description: D is ample if and only if the corresponding invertible sheaf $L = \mathcal{O}_X(D)$ has the property that for each coherent sheaf F on X, $F \otimes L^n$ is generated by global sections for $n \gg 0$. One can think of this condition as generalizing the property of a linear system having no base points. Indeed, a linear system $|D|$ has no base points if and only if the sheaf $L = \mathcal{O}_X(D)$ itself is generated by global sections.

The second description is a cohomological one: D is ample if and only if for every coherent sheaf F on X, the cohomology groups $H^i(X, F \otimes L^n)$ are zero for $i > 0$ and n sufficiently large. This generalizes Serre's theorem on the vanishing of cohomology of twisted sheaves on projective space.

In the latter part of the chapter, we are mostly concerned with numerical characterizations of ampleness. That is to say, we wish to characterize ample divisors by their intersection properties with other subvarieties. Since any two subvarieties of projective space, of complementary dimension, have a non-empty intersection, we see immediately that a hyperplane section of a projective variety X must meet every curve in X. This implies that an ample divisor D on X will have a positive intersection number with every curve on X (we say that D is numerically positive). So one can ask conversely, is every numerically positive divisor ample? The answer is no (at least in characteristic zero), even if D is effective, as is shown by examples of Mumford and Ramanujam, given in §10.

However, by strengthening this condition suitably, one can obtain numerical criteria of ampleness. One way is to require that D and all its self-intersections be numerically positive. This is the Nakai-Moisezon criterion, proved in §5. It says that D is ample if and only if $(D^s.Y) > 0$ for all subvarieties Y of dimension s, and for all $s \leq \dim X$.

Another way of strengthening the condition of numerical positivity is based on the following observation. Again suppose D is a hyperplane section of a projective variety X. Then for any curve $Y \subseteq X$, (D.Y) is just the degree of the curve Y, considered as a curve in projective space. This suggests that in the general case we should establish some method of measuring the size of the curve Y, and require that (D.Y) be large for large Y.

For the criterion of Seshadri, we define m(Y) to be the maximum of the multiplicities of the points of Y. His theorem says that D is ample if and only if there is an $\epsilon > 0$ such that for all curves Y on X, $(D.Y) \geq \epsilon.m(Y)$. This theorem has not been previously published.

A similar result is the theorem of Kleiman, which uses a norm $\| \;\|$ on the real vector space $A_1(X)\otimes_{\mathbb{Z}}\mathbb{R}$, where $A_1(X)$ is the group of 1-cycles on X modulo numerical equivalence. His theorem, for which we give a new proof in §8, says that D is ample if and only if there is an $\epsilon > 0$ such that $(D.Y) \geq \epsilon\|Y\|$ for all curves Y on X.

One important application of ample divisors is to proving that certain varieties are projective. It is known that every non-singular curve or surface is projective (see Chapter II, §4). On the other hand, there are non-singular complete three-folds which are not projective (see for example Hironaka [2]). So it is of interest to give criteria for varieties to be projective. In §9, we give Kleiman's proof of Chevalley's conjecture, which says that X is projective if and only if every finite set of points is contained in an affine open set.

In the Appendix, §10, we give a new proof (also the first published proof) of the example, due to Mumford, of a divisor which is numerically positive but not ample. It depends on the existence of a stable vector bundle (on a curve) all of whose symmetric powers are also stable. Also we give a new example, due to Ramanujam, of an effective divisor which is numerically positive but not ample.

§0. Generalities on Divisors.

Let X be a non-singular variety (defined over an algebraically closed field k as always). A closed irreducible subvariety Y of codimension 1 in X is called a prime divisor on X. An element D of the free abelian group generated by the set of prime divisors Y_i is called a divisor on X. If $D = \sum_i n_i Y_i$, $n_i \neq 0$, is a divisor on X, we define the support of D, denoted Supp (D), to be $\bigcup_i Y_i$.

Let Y be a prime divisor on X. At any point $y \in Y$, Y determines a prime ideal $\mathfrak{p}$ of height one in the local ring $\mathcal{O}_{X,y}$. Since X is non-singular, $\mathcal{O}_{X,y}$ is regular, hence UFD, so $\mathfrak{p}$ is principal, say $\mathfrak{p} = (f)$. Thus Y is defined by the equation $f = 0$ at y, and hence in a neighborhood of y. Therefore there exists an affine open covering $\{U_\alpha\}$ of X and elements $f_\alpha \in \Gamma(U_\alpha, \mathcal{O}_X)$ such that

i) for each α, the closed subscheme $Y \cap U_\alpha$ of U_α is defined by the principal ideal (f_α), and

ii) for all α,β; f_α/f_β is a unit in $\Gamma(U_\alpha \cap U_\beta, \mathcal{O}_X)$.

Suppose now $D = \sum_i n_i Y_i$ is any divisor on X. Then we can find an affine open covering $\{U_\alpha\}$ of X and elements $f_{i,\alpha} \in \Gamma(U_\alpha, \mathcal{O}_X)$ such that each Y_i corresponds to $\{(U_\alpha, f_{i,\alpha})\}$. The elements $f_\alpha = \prod_i f_{i,\alpha}^{n_i} \in K(X)$ (= the field of rational functions on X) are such that, for all α,β; f_α/f_β is a unit in $\Gamma(U_\alpha \cap U_\beta, \mathcal{O}_X)$. Thus the divisor D determines the system $\{(U_\alpha, f_\alpha)\}$.

Such a description for divisors is no longer possible if X is not non-singular. However, borrowing the description, we make suitable definition for a divisor on any scheme which would coincide with the above if the scheme happens to be non-singular.

§1. Cartier Divisors.

We assume that X is a noetherian scheme.

Let $\mathcal{K}_X$ be the sheaf of total quotient rings of $\mathcal{O}_X$, i.e., for affine open $U \subseteq X$, $\Gamma(U, \mathcal{K}_X)$ is the total quotient ring of $\Gamma(U, \mathcal{O}_X)$. Let $\mathcal{K}_X^*$ be the sheaf of groups of units of the sheaf of rings $\mathcal{K}_X$. Note that we have $\mathcal{O}_X \subseteq \mathcal{K}_X$ and $\mathcal{O}_X^* \subseteq \mathcal{K}_X^*$.

A Cartier divisor (or simply a divisor) D on X is represented by a collection $\{(U_i, f_i)\}$ where $\{U_i\}$ is an affine open covering of X, and $f_i \in \Gamma(U_i, \mathcal{K}_X^*)$, such that for all i,j, $f_i/f_j \in \Gamma(U_i \cap U_j, \mathcal{O}_X^*)$.

Two such collections $\{(U_i, f_i)\}$ and $\{(V_j, g_j)\}$ define the same divisor D if there exists a common refinement $\mathcal{W} = \{W_k\}$ of $\mathcal{U} = \{U_i\}$ and $\mathcal{V} = \{V_j\}$ by affine open $W_k \subseteq X$ such that the restrictions of f_i and g_j to smaller open sets differ by a unit. In other words, noticing that a representative $\{(U_i, f_i)\}$ defines an element of $\check{H}^1(\mathcal{U}, \mathcal{O}_X^*)$, we demand that the canonical images of $\{(U_i, f_i)\}$ and $\{(V_g, g_j)\}$ in $\check{H}^1(\mathcal{W}, \mathcal{O}_X^*)$ coincide.

The elements $f_i \in \Gamma(U_i, \mathcal{K}_X^*)$ are called the local equations of D, and they are unique up to a unit in $\mathcal{O}_X$. Thus a Cartier divisor D is just an element of $H^0(X, \mathcal{K}_X^*/\mathcal{O}_X^*)$.

The support of a divisor D, denoted Supp (D), is defined as the closed subset $\{x \in X \mid 1$ is not a local equation for D at $x\}$. We sometimes write $x \in D$ if $x \in$ Supp (D).

The set of all Cartier divisors on X, denoted Div (X), is a group under addition: suppose $D_1 = \{(U_i, f_i)\}$ and $D_2 = \{(U_i, g_i)\}$ are two divisors, then we define D_1+D_2 to be the divisor $\{(U_i, f_i g_i)\}$.

Given a divisor $D = \{(U_i, f_i)\}$ on X, we associate with it an invertible sheaf $\mathcal{O}_X(D)$ of $\mathcal{O}_X$-modules in a natural way: namely $\mathcal{O}_X(D)$ is the subsheaf of $\mathcal{K}_X$ generated locally by f_i^{-1} on U_i. Or equivalently, $\mathcal{O}_X(D)$ is constructed by starting with $\mathcal{O}_{U_i}$ on U_i, and patching them on $U_i \cap U_j$ via f_i/f_j.

We recall that Pic (X), the <u>Picard group</u> of X, is the group of isomorphism classes of invertible sheaves on X under $\otimes$. It is canonically isomorphic to $H^1(X, \mathcal{O}_X^*)$.

We have a canonical homomorphism δ: Div (X) $\longrightarrow$ Pic (X) given by $D \longmapsto$ the isomorphism class of $\mathcal{O}_X(D)$. What is $\ker \delta$?

A divisor D is called <u>principal</u> if there exists an $f \in H^o(X, \mathcal{K}_X^*)$ such that f is the local equation of D at every $x \in X$. Such a divisor is denoted by (f).

Two divisors D_1 and D_2 are said to be <u>linearly equivalent</u> if $D_1 - D_2$ is principal. In this case we write $D_1 \sim D_2$.

The set of all divisors linearly equivalent to a divisor D is called the <u>divisor class</u> of D.

The following proposition gives $\ker \delta$:

<u>Proposition 1.1</u>. A divisor D is principal $\Longleftrightarrow \mathcal{O}_X(D) \cong \mathcal{O}_X$ as sheaves of $\mathcal{O}_X$-modules.

<u>Proof</u>. We have an exact sequence of sheaves on X

$$0 \longrightarrow \mathcal{O}_X^* \longrightarrow \mathcal{K}_X^* \longrightarrow \mathcal{K}_X^*/\mathcal{O}_X^* \longrightarrow 0 .$$

This gives the cohomology exact sequence

$$0 \longrightarrow H^o(X, \mathcal{O}_X^*) \longrightarrow H^o(X, \mathcal{K}_X^*) \xrightarrow{\alpha} H^o(X, \mathcal{K}_X^*/\mathcal{O}_X^*)$$
$$\xrightarrow{\delta} H^1(X, \mathcal{O}_X^*) \longrightarrow H^1(X, \mathcal{K}_X^*) .$$

Hence $\ker \delta = \operatorname{Im} \alpha$ = Principal divisors.

Proposition 1.2. If X is an integral scheme, the map δ is surjective.

Proof. Since X is integral, we have $\mathcal{K}_X = K(X)$, the constant sheaf of the field of rational functions on X. But then $\mathcal{K}_X^*$ is also the constant sheaf of the group $(K(X))^*$. Now, since any constant sheaf on an irreducible scheme is flasque, we have $H^1(X, \mathcal{K}_X^*) = 0$. Hence δ is surjective.

Remark. Nakai has shown that for a projective scheme X, the map δ: Div $(X) \longrightarrow$ Pic (X) is surjective (see Nakai [3], Theorem 4, p. 301).

On the other hand, the following example shows that δ is not always surjective.

Example 1.3. (Kleiman) There is a complete scheme X' and an invertible sheaf L' on X', such that L' is not isomorphic to $\mathcal{O}_{X'}(D')$ for any Cartier divisor D' on X'.

Let X be a complete non-singular, non-projective 3-fold containing two disjoint irreducible curves C_1 and C_2 such that C_1+C_2 is algebraically equivalent to zero. (Such an X exists by Hironaka [1] or [2]. For example, take $X_o = \mathbb{P}^3$ and take two non-singular curves Y and Z in X_o which meet transversally at two points P and Q. Near P, blow up first Y and then Z. Near Q, blow up first Z and then Y. Elsewhere blow up Y and Z. Then the new variety thus obtained has the above properties).

If D is any prime divisor which intersects C_1 properly

(and there are such, since X is non-singular), then D must contain C_2, because $(D\cdot C_2) = -(D\cdot C_1) < 0$.

Let P be a point of C_2. Define a new scheme X', which has the same topological space as X, and the same structure sheaf everywhere except P. At P, define $\mathcal{O}_{X',P} = \mathcal{O}_{X,P} \oplus k$, with multiplication defined by $(a,\lambda)(b,\mu) = (ab, a\mu+b\lambda)$. Then the only non-zero divisors of $\mathcal{O}_{X',P}$ are units, so no Cartier divisor on X' can contain P. In fact, one sees easily that there is a one-to-one correspondence

$$\mathrm{Div}\,(X') = \{D \in \mathrm{Div}\,(X) \mid P \notin D\}\,.$$

On the other hand, we have an exact sequence

$$0 \longrightarrow k \longrightarrow \mathcal{O}_{X'}^{*} \longrightarrow \mathcal{O}_{X}^{*} \longrightarrow 0$$

so $\mathrm{Pic}\,(X') \approx \mathrm{Pic}\,(X)$.

Let D be a divisor on X which intersects C_1 properly. Let $L = \mathcal{O}_X(D)$, and let L' be the corresponding invertible sheaf on X'. Then L' is not isomorphic to $\mathcal{O}_{X'}(D')$ for any Cartier divisor D' on X'.

<u>Exercise 1.4</u>. Let X be the affine plane curve given by the equation $y^2 = x^3$. In other words $X = \mathrm{Spec}(k[x,y]/(y^2-x^3))$. Let P be the point $(0,0)$. Let $\mathrm{Div}_P X$ be the group of Cartier divisors on X having support contained in P. Show that $\mathrm{Div}_P X \cong k^{+} \oplus \mathbb{Z}$, where k^{+} is the additive group of the field.

<u>Exercise 1.5</u>. Let X be the plane curve given by $y^2 = x^2+x^3$, and again let $P = (0,0)$. Show in this case that $\mathrm{Div}_P X \cong k^{*} \oplus \mathbb{Z} \oplus \mathbb{Z}$, where k^{*} is the multiplicative group of the field.

These exercises show that the group of Cartier divisors on a singular scheme need not be a free abelian group, as it is in the case of a non-singular variety.

§2. Linear systems.

In this section we discuss effective Cartier divisors, linear systems, and associated maps into projective space.

A Cartier divisor $D = \{U_i, f_i\}$ is effective, written $D \geq 0$, if f_i is a non-zero divisor in $\Gamma(U_i, \mathcal{O}_{U_i})$ for each i.

Proposition 2.1. There are natural one-to-one correspondences between the following three sets:

A = {effective Cartier divisors D},

B = {invertible sheaves L on X, together with a section $s \in H^0(X,L)$, which is locally a non-zero divisor, and which is unique to within a unit $\alpha \in H^0(X, \mathcal{O}_X^*)$},

C = {subschemes $Y \subseteq X$, locally defined by a single equation}.

Proof. We define maps as follows:

$A \longrightarrow B$. Given an effective Cartier divisor D, let $L = \mathcal{O}_X(D)$. Then by construction, $\mathcal{O}_X \subseteq \mathcal{O}_X(D)$. Hence the global section $1 \in H^0(X, \mathcal{O}_X)$ defines a section $s \in H^0(X, \mathcal{O}_X(D))$.

$B \longrightarrow C$. Given an invertible sheaf L, and a global section $s \in H^0(X,L)$ which is locally a non-zero-divisor, we define the subscheme Y as follows. On a small open set U, pick an isomorphism $L|_U \cong \mathcal{O}_U$. Then transport s to obtain a non-zero-divisor $f \in H^0(U, \mathcal{O}_U)$. Let Y be the subscheme defined by f. We say Y is the subscheme of zeros of the section s.

$C \longrightarrow A$. Given Y, let f_i be a local equation for Y on an open set U_i. Then take $D = \{U_i, f_i\}$.

We leave to the reader the verification that these maps induce one-to-one correspondences between the three sets A, B, C.

<u>Exercise 2.2.</u> Under the above correspondence, show that the sheaf of ideals of the subscheme Y is just $\mathcal{O}_X(-D)$.

Having established the correspondences above, we will often identify the sets A and C, and we will write D for the subscheme defined by the Cartier divisor D.

If D is any divisor, we define the <u>complete linear system</u>, $|D|$, to be the set of all effective divisors D' which are linearly equivalent to D. Thus the set $|D|$ is in one-to-one correspondence with the set of sections $s \in H^o(X,\mathcal{O}_X(D))$ which are locally non-zero-divisors, modulo units $\alpha \in H^o(X,\mathcal{O}_X^*)$.

<u>We will now assume, for simplicity, that</u> X <u>is a complete integral scheme over</u> k.

In that case $H^o(X,\mathcal{O}_X^*) = k^*$, and $H^o(X,\mathcal{O}_X(D))$ is finite-dimensional. Thus $|D|$ is in one-to-one correspondence with the projective space $H^o(X,\mathcal{O}_X(D))/k^*$. This justifies defining the <u>dimension</u> of $|D|$ as follows:

$$\dim |D| = \dim H^o(X,\mathcal{O}_X(D)) - 1.$$

A <u>linear system</u> on X is a set $\mathcal{D}$ of effective divisors such that

i) $D, D' \in \mathcal{D} \Longrightarrow D \sim D'$ (so that in particular $\mathcal{D} \subseteq |D|$ for some D).

ii) there exists a subspace $V \subseteq H^o(X,\mathcal{O}_X(D))$ such that

$\mathfrak{D} = \{(s)_o | s \in V\}$, where $(s)_o$ denotes the zeros of the section s.

The <u>dimension of a linear system</u> $\mathfrak{D}$ is defined by $\dim \mathfrak{D} = \dim V - 1$.

We say that a point $x \in X$ is a <u>base point</u> for a linear system $\mathfrak{D}$ if $x \in D$ for every $D \in \mathfrak{D}$. If $\mathfrak{D}$ has no base points, we say $\mathfrak{D}$ is <u>base point free</u>.

<u>Proposition 2.3</u>. A complete linear system $|D|$ is base point free $\Longleftrightarrow$ $\mathcal{O}_X(D)$ is generated by its global sections.

<u>Proof</u>. By definition, $|D|$ has no base points

$\Longleftrightarrow \forall x \in X, \exists D' \in |D|$ such that $x \notin D'$

$\Longleftrightarrow s'(x) \neq 0$ where $D' = (s')_o$, $s' \in H^o(X, \mathcal{O}_X(D))$.

$\Longleftrightarrow s'$ generates $\mathcal{O}_X(D)$ at x.

Now we shall study the relation between linear systems on X and rational maps of X into the projective space $\mathbb{P}^n_k$.

Let $\mathfrak{D} \subseteq |D|$ be a linear system on X, and let $L = \mathcal{O}_X(D)$. Let $V \subseteq H^o(X,L)$ be the subspace associated to $\mathfrak{D}$. Then given sections $s_o,\ldots,s_n \in V$, which span V, we can define a rational map $\varphi: X \longrightarrow \mathbb{P}^n_k$ as follows.

Let $U = X - \{\text{base points of } \mathfrak{D}\}$. For a closed point $x \in U$, define $\varphi(x) = (s_o(x),\ldots,s_n(x))$, where $s_i(x)$ is the image of s_i in $L_x/\mathfrak{m}_x L_x$. Since the s_i's span V, we have at least one $s_i(x) \neq 0$. Furthermore, since $L_x/\mathfrak{m}_x L_x \cong k$, $\varphi(x)$ gives a well-defined point in the projective space $\mathbb{P}^n_k$.

We check that this is a morphism and $L|_U \cong \varphi^*(\mathcal{O}_{\mathbb{P}}(1))$.

Moreover we have $s_i = \varphi^*(X_i)$ where X_i's are the canonical sections of $\mathcal{O}_{\mathbb{P}}(1)$.

For a more scheme-theoretic version of this definition, see [EGA, II, §4].

Note that φ is a morphism if $\mathfrak{D}$ is base point free. Conversely, given a morphism $\varphi: X \longrightarrow \mathbb{P}^n_k$, we have an invertible sheaf $L = \varphi^*(\mathcal{O}_{\mathbb{P}}(1))$ and sections $s_i = \varphi^*(X_i)$, $0 \leq i \leq n$, such that for every $x \in X$, at least one $s_i(x) \neq 0$. If V is the subspace generated by the s_i's, we have a linear system $\mathfrak{D} = \{(s)_0 | s \in V\}$ without base points. Further the morphism defined by $\mathfrak{D}$ and $s_0, \ldots, s_n$ is φ. Thus we have established

Proposition 2.4. Let X be a complete integral scheme over k. Then there is a natural one-to-one correspondence between morphisms of X into $\mathbb{P}^n_k$ and linear systems $\mathfrak{D}$ without base points, together with $n+1$ sections $s_0, \ldots, s_n \in V$, which span V (where $V \subseteq H^0(X, \mathcal{O}_X(D))$ is the subspace associated with $\mathfrak{D} \subseteq |D|$).

Let $x \in X$. We recall that the Zariski tangent space at x, denoted T_x, is the vector space $T_x = \mathrm{Hom}_{k(x)}(\mathfrak{m}_x/\mathfrak{m}_x^2, k(x))$.

We say that a complete linear system $|D|$ separates points of X if given $x, y \in X$, $x \neq y$, there is a $D' \in |D|$ such that $x \in D'$ and $y \notin D'$. Further we say $|D|$ separates infinitely near points of X if for every closed point $x \in X$ and every non-zero tangent vector $t \in T_x$, there is a $D' \in |D|$ such that $x \in D'$ and $t \notin D'$, i.e., if f is a local equation of D' at x (so that $f_x \in \mathfrak{m}_x$), we want $t(\overline{f}_x) \neq 0$ where $\overline{f}_x$ is the residue of f_x mod $\mathfrak{m}_x^2$).

We now give a criterion for the morphism φ defined by a linear system $\mathcal{D}$ to be a closed immersion.

Theorem 2.5. Let X be a complete integral scheme. Let $|D|$ be a complete linear system without base points. Let $s_o,\ldots,s_n \in H^o(X,\mathcal{O}_X(D))$ be sections which span, and let φ be the associated morphism of X into $\mathbb{P}^n_k$. Then φ is a closed immersion $\Longleftrightarrow$ $|D|$ separates points and also separates infinitely near points.

Proof. Suppose $\varphi: X \longrightarrow \mathbb{P}^n_k$ is a closed immersion. Then we can assume X is a closed subscheme of $\mathbb{P} = \mathbb{P}^n_k$ via φ. We have $L = \mathcal{O}_{\mathbb{P}}(1)\big|_X$ and $|D| = \{X \cap H \big| X \nsubseteq H = \text{hyperplane in } \mathbb{P}\}$. Now given $x,y \in X$, $x \neq y$, we can certainly find a hyperplane H such that $x \in H$ and $y \notin H$. Thus $|D|$ separates points of X. That $|D|$ separates infinitely near points becomes obvious once we observe that the vector space $\mathfrak{m}_x/\mathfrak{m}_x^2$, x closed point of X, is generated by (homogeneous) linear forms.

Conversely, suppose $|D|$ separates points and infinitely near points of X. Then we claim that φ is injective. For, let $x,y \in X$ and $x \neq y$. We have a $D' \in |D|$ such that $x \in D'$ and $y \notin D'$. Let $s' \in H^o(X,L)$ be a section whose divisor of zeros is D'. We have $s'(x) = 0$ and $s'(y) \neq 0$. But $s' = \sum_{i=0}^{n} \lambda_i s_i$, $\lambda_i \in k$. Hence $s'(x) = \sum \lambda_i s_i(x)$ and $s'(y) = \sum \lambda_i s_i(y)$ implying that

$$\varphi(x) = (s_i(x)) \neq (s_i(y)) = \varphi(y).$$

Let I be the kernel of the homomorphism $\mathcal{O}_{\mathbb{P}} \longrightarrow \varphi_*(\mathcal{O}_X)$. Let X' be the closed subscheme of $\mathbb{P}$ defined by the ideal I. We want to prove that $\varphi: X \longrightarrow X'$ is an isomorphism.

By construction, the homomorphism $\mathcal{O}_{X'} \longrightarrow \varphi_*(\mathcal{O}_X)$ is injective. Let $x \in X$ be a closed point of X and $x' = \varphi(x)$. Since $|D|$ separates infinitely near points of X, the induced map $T_x \longrightarrow T_{x'}$ is injective, i.e., the linear map $\mathfrak{m}_{x'}/\mathfrak{m}_{x'}^2 \longrightarrow \mathfrak{m}_x/\mathfrak{m}_x^2$ is surjective. Finally, since X is proper over k, it is proper over $\mathbb{P}$, i.e., φ is a proper morphism. Hence by [EGA, III, 3.2.2], $\mathcal{O}_X$ is a coherent $\mathcal{O}_{X'}$-module. In particular $\mathcal{O}_x$ is a finite $\mathcal{O}_{x'}$-module. Now it follows that φ is an isomorphism in view of the following

Lemma 2.6. Let $f: A \longrightarrow B$ be a local homomorphism of noetherian local rings such that

1) f is injective,
2) $\mathfrak{m}_A/\mathfrak{m}_A^2 \longrightarrow \mathfrak{m}_B/\mathfrak{m}_B^2$ is surjective,
3) B is a finite A-module, and
4) $k(A) = k(B)$ (= residue fields of A and B).

Then f is an isomorphism.

Proof. Because of (2) and Nakayama on B, we get $\mathfrak{m}_A.B = \mathfrak{m}_B$. Now applying Nakayama on A using (3) and (4), we get f is surjective and hence an isomorphism.

§3. Ample Divisors.

In this section we define ample divisors, and we prove a theorem of Serre and Grothendieck, characterizing them in terms of sheaf-theoretic and cohomological properties.

We use the word "complete scheme" as a synonym for a "scheme proper over k", and the word "line bundle" for an "invertible sheaf". For an invertible sheaf F of $\mathcal{O}_X$-modules, we use the notation F^n for the n-fold tensor product $F^{\otimes n}$. We have for example $\mathcal{O}_X(nD) = (\mathcal{O}_X(D))^n$ for every $D \in \mathrm{Div}(X)$ and $n \in \mathbb{Z}$.

Let X be a complete scheme. A divisor D on X is said to be very ample if there exists a closed immersion $X \hookrightarrow \mathbb{P} = \mathbb{P}^r_k$ (for some r) such that

$$\mathcal{O}_X(D) \cong \mathcal{O}_{\mathbb{P}}(1)\big|_X .$$

A divisor D is said to be ample if nD is very ample for some positive integer n. We apply the same terminology to the associated line bundle $L = \mathcal{O}_X(D)$.

Note that the existence of an ample divisor on X implies in particular X is a projective scheme.

It is clear that a very ample line bundle is generated by its global sections. The following is the first fundamental theorem which gives a cohomological characterization for a divisor to be ample.

Theorem 3.1. (Serre-Grothendieck [EGA, III, 2.6.1]) Let X be a complete integral scheme, and let L be a line bundle on X. Then the following are equivalent:

i) L is ample,

ii) For every coherent sheaf F on X, we have

$$H^i(X, F \otimes L^n) = 0 \quad \text{for all } i > 0, \text{ and } n \gg 0,$$

iii) For every coherent sheaf F on X, the sheaf $F \otimes L^n$ is generated by its global sections for $n \gg 0$.

Proof. (i) $\Longrightarrow$ (ii):

We can assume without loss of generality that L is very ample. For, by (i), L^s is very ample for some $s > 0$. Suppose we have proved (ii) for L^s. Then given a coherent sheaf F on X, we can find for every $r = 0,\ldots,s-1$, an integer $n_r > 0$ such that for all $n \geq n_r$, we have $H^i(X, (F \otimes L^r) \otimes L^{sn}) = 0$ for all $i > 0$. If we choose $N \geq \max_r (sn_r)$, we have for all $n \geq N$,

$H^i(X, F \otimes L^n) = H^i(X, (F \otimes L^r) \otimes L^{s(n_r+k)}) = 0$ for all $i > 0$.

Thus we can replace L by L^s.

Now L being very ample, we have a closed immersion $X \hookrightarrow \mathbb{P}^r_k$ (for some r) such that $L = \mathcal{O}_{\mathbb{P}}(1)|_X$. But then $F \otimes L^n = F(n) = F \otimes \mathcal{O}_{\mathbb{P}}(n)$ and hence by [EGA, III, 2.2.1], we have $H^i(X,F(n)) = 0$ for all $i > 0$ and $n \gg 0$.

(ii) $\Longrightarrow$ (iii):

Let F be a given coherent sheaf on X. Let $x \in X$ be a closed point, and let I_x be the sheaf of ideals of $\mathcal{O}_X$ defining the closed subscheme $\{x\}$. The exact sequence

$$0 \longrightarrow I_xF \longrightarrow F \longrightarrow F/I_xF \longrightarrow 0$$

gives an exact sequence

$$0 \longrightarrow I_xF \otimes L^n \longrightarrow F \otimes L^n \longrightarrow F/I_xF \otimes L^n \longrightarrow 0$$

and hence the cohomology exact sequence

$$H^o(X, F\otimes L^n) \longrightarrow H^o(X, F/I_x F\otimes L^n) \longrightarrow H^1(X, I_x F\otimes L^n) .$$

But $I_x F$ is coherent on X and so by (ii), we have $H^1(X, I_x F\otimes L^n) = 0$ for $n \gg 0$, say for all $n \geq n_o$. Therefore $H^o(X, F\otimes L^n) \longrightarrow H^o(X, F/I_x F\otimes L^n)$ is surjective for all $n \geq n_o$. It follows by Nakayama that $F \otimes L^n$ is generated by its global sections at x (and hence in a neighborhood) for all $n \geq n_o$.

In particular, taking $F = \mathcal{O}_X$, we find that L^{n_1} is generated by its global sections in a neighborhood U of x for some $n_1 > 0$. Thus for each $r = 0,1,\ldots,n_1$, there is a neighborhood U_r of x such that the sheaf $F \otimes L^{n_o+r}$ is generated by its global sections in U_r. Let

$$U_x = U \cap U_o \cap \ldots \cap U_{n_1} .$$

Then for any $m > 0$, the sheaf $F \otimes L^{n_o+r} \otimes L^{n_1 m}$ is generated by its global sections in U_x. But any large enough n can be written in the form $n_o + r + n_1 m$. Hence $F \otimes L^n$ is generated by its global sections in U_x for $n \gg 0$ (say for $n \geq n(x)$).

We repeat this process for each closed point $x \in X$. By quasi-compacity we can cover X with a finite number of these open sets, say $U_{x_1},\ldots,U_{x_t}$. Take $N = \max_i (n(x_i))$. Then for all $n \geq N$, $F \otimes L^n$ is generated by its global sections. This proves (iii).

(iii) $\Longrightarrow$ (i):

By (iii), L^s is generated by its global sections for some $s > 0$. Therefore replacing L by L^s, we can assume that L is generated by its global sections. Hence L^n is generated by its global sections for every $n > 0$.

Let $\varphi_n : X \longrightarrow \mathbb{P}_k^{N_n}$ (for some N_n) be the morphism defined by L^n. To prove that L^n is very ample for some n, we have only to check by Theorem 2.5, that φ_n separates points and infinitely near points.

The following technique gives this: Let $x \in X$ be a closed point, and let I_x be the sheaf of ideals defining the closed subscheme $\{x\}$. Then we claim that φ_n is an isomorphism at $x \Longleftrightarrow I_x \otimes L^n$ is generated by its global sections. The implication $\Longrightarrow$ is obvious. Conversely, let $s \in H^0(X, I_x \otimes L^n)$. Then s, as an element of $H^0(X, L^n)$, is such that $s(x) = 0$. On the other hand, if $y \neq x$, $y \in X$, we can choose s such that $s(y) \neq 0$. Thus φ_n separates x from every other point of X. Similarly we conclude that φ_n separates x from its infinitely near points (since $I_x/I_x^2 \otimes L^n$ is also generated by its global sections). Hence in view of the proof of Theorem 2.5, it follows that φ_n is an isomorphism at x.

<u>Proof of the theorem</u>, <u>continued</u>.

By (iii), for any point x, $I_x \otimes L^n$ is generated by its global sections for $n \gg 0$. Therefore φ_n is an isomorphism at x for all $n \gg 0$. For any n, let

$$U_n = \{y \in X \mid \varphi_n \text{ is an isomorphism at } y\}.$$

By [EGA, I, 6.5.4], this is an open set. We claim that for all n we have

$$U_n \subseteq U_{n+1} \subseteq \dots .$$

For, let $y \in U_n$. Then φ_n is an isomorphism at y and hence

$I_y \otimes L^n$ is generated by its global sections. But L is generated by its global sections. Hence $I_y \otimes L^n \otimes L$ is generated by its global sections. But then φ_{n+1} is an isomorphism at y, i.e., $y \in U_{n+1}$.

Since for every $x \in X$, we have an n such that $x \in U_n$, $\{U_n\}$ is an open covering of X, and hence by quasi-compacity we have $X = U_n$ for some $n \gg 0$. Thus for sufficiently large n, φ_n is a closed immersion. This completes the proof of the theorem.

Remark. The above theorem remains true even when X is not necessarily integral, i.e., when X is any complete scheme. We leave the proof as an exercise. In the sequel we use this theorem in this generality without further comment.

Exercise 3.2. Suppose X is a non-singular complete curve of genus g. Use the Riemann-Roch theorem for a divisor D on X and prove that

1) $\deg(D) \geq g \Longrightarrow |D| \neq \emptyset$.

2) $\deg(D) \geq 2g \Longrightarrow |D|$ has no base points.

3) $\deg(D) \geq 2g+1 \Longrightarrow D$ is very ample.

4) $\deg(D) > 0 \Longleftrightarrow D$ is ample.

§4. Functorial Properties.

We study the behavior of ample divisors under base change. Throughout this section X will denote a complete scheme.

Proposition 4.1. Let L be an ample line bundle on X. Then for every closed subscheme Y of X, $L|_Y$ is ample.

Proof. Let $L_Y = L|_Y$. We have $L_Y^n = L^n|_Y$. Suppose a coherent sheaf F on Y is given. We have $H^i(Y, F\otimes L_Y^n) = H^i(X, \bar{F}\otimes L^n)$ where $\bar{F}$ is the sheaf on X extending F with zero outside Y. By Theorem 3.1 (ii), we have $H^i(X, \bar{F}\otimes L^n) = 0$ for all $i > 0$ and $n >> 0$ and hence L_Y is ample on Y.

Proposition 4.2. Let L be a line bundle on X. Then L is ample on $X \Longleftrightarrow L_{red}$ is ample on X_{red}.

Proof. Since X_{red} is closed in X, we have only to prove $\Longleftarrow$:

Let F be a coherent sheaf on X. Let N be the nil radical of $\mathcal{O}_X$. We have $N^r = 0$ for some $r > 0$ since X is noetherian. Consider the filtration

$$F \supset NF \supset \dots \supset N^rF = (0) .$$

For each $i = 1,\dots,r-1$, we have an exact sequence

$$0 \longrightarrow N^iF \longrightarrow N^{i-1}F \longrightarrow N^{i-1}F/N^iF \longrightarrow 0$$

and hence the cohomology exact sequence

$$\longrightarrow H^p(X,N^iF\otimes L^n) \longrightarrow H^p(X,N^{i-1}F\otimes L^n) \longrightarrow H^p(X,N^{i-1}F/N^iF \otimes L^n)$$

$$\longrightarrow H^{p+1}(X, N^iF\otimes L^n) \longrightarrow \cdots .$$

We want to show that $H^p(X, F\otimes L^n) = 0$ for all $p > 0$ and $n \gg 0$. This we achieve by descending induction on i. Note that $N^{i-1}F/N^iF$ is a coherent $\mathcal{O}_{X_{red}}$-module for every i. Hence by hypothesis we have $H^p(X,N^{i-1}F/N^iF\otimes L^n) = 0$ for all $p > 0$ and $n \gg 0$. Since $N^i = 0$ for $i \geq r$, we have $H^p(X,N^iF\otimes L^n) = 0$ for all $p > 0$, $i \geq r$ and $n \gg 0$. The cohomology exact sequences together with the induction hypothesis give that $H^p(X,F\otimes L^n) = 0$ for all $p > 0$ and $n \gg 0$.

<u>Proposition 4.3</u>. A line bundle L on X is ample $\iff$ $L|_{X_i}$ is ample on X_i for every irreducible component X_i of X.

<u>Proof</u>. Each irreducible component of X being closed, we have only to prove $\Longleftarrow$:

By the above proposition, we can assume X is reduced. Let $X = \bigcup_{i=1}^{r} X_i$, X_i irreducible components of X. Let I_i be the ideal of X_i. We prove the result by induction on r. If $r = 1$, there is nothing to prove. Assume the induction hypothesis. Consider for example the exact sequence

$$0 \longrightarrow I_1F \longrightarrow F \longrightarrow F/I_1F \longrightarrow 0$$

where F is any coherent sheaf on X. We have the cohomology exact sequence

$$\longrightarrow H^p(X,I_1F\otimes L^n) \longrightarrow H^p(X,F\otimes L^n) \longrightarrow H^p(X,F/I_1F\otimes L^n)$$
$$\longrightarrow H^{p+1}(X,I_1F\otimes L^n) \longrightarrow \cdots .$$

By hypothesis, we have $H^p(X, F/I_1F \otimes L^n) = 0$ for all $p > 0$ and $n \gg 0$. But the induction hypothesis gives $H^p(X,I_1F\otimes L^n) = 0$ for all $p > 0$ and $n \gg 0$. Since $\mathrm{Supp}(I_1F) \subseteq X_2 \cup \ldots \cup X_r$. Hence we have $H^p(X,F\otimes L^n) = 0$ for all $p > 0$ and $n >> 0$.

Proposition 4.4. Let $f: X \longrightarrow Y$ be a finite surjective morphism, and let L be a line bundle on Y. Then f^*L is ample on $X \Longleftrightarrow L$ is ample on Y.

Proof. Suppose L is ample on Y. Suppose f is a finite morphism (not necessarily surjective). Let F be a coherent sheaf on X. Note that $f_*(F)$ is a coherent $\mathcal{O}_Y$-module. Also we have the projection formula

$$f_*(F \otimes (f^*L)^n) = f_*(F) \otimes L^n \text{ [EGA, 0, 5.4.6].}$$

Since f is finite (hence affine), by [EGA, III, 1.3.3], we have $H^i(Y,f_*(f^*L)^n) \xrightarrow{\sim} H^i(X,f^*L)^n)$ for all i. Thus

$$\begin{array}{ccc} H^i(X,F\otimes(f^*L)^n) & \xleftarrow{\sim} & H^i(Y,f_*(F\otimes(f^*L)^n)) \\ & & \| \\ & & H^i(Y,f_*(F)\otimes L^n) \end{array} .$$

But then $H^i(Y,f_*(F)\otimes L^n) = 0$ for all $i > 0$ and $n >> 0$. This implies that f^*L is ample on X.

Conversely, suppose f^*L is ample on X. We may assume without loss of generality that both X and Y are integral. We deduce the result from the

Lemma 4.5. Let $f: X \longrightarrow Y$ be a finite surjective morphism of degree m of the integral schemes X and Y. Then for every coherent sheaf F on Y, there is a coherent sheaf G on X and a homomorphism $u: f_*(G) \longrightarrow F^{\oplus m}$ such that u is a generic isomorphism (i.e., an isomorphism in a neighborhood of the generic point of Y).

Proof of the proposition. Let F be a coherent sheaf on Y. We will use noetherian induction on the Supp (F). Let G be as given in the above lemma. Let K and C be the kernel and the cokernel of $u: f_*(G) \longrightarrow F^{\oplus m}$ respectively. We have the exact sequences

$$0 \longrightarrow K \longrightarrow f_*(G) \longrightarrow \operatorname{Im} u \longrightarrow 0$$

$$0 \longrightarrow \operatorname{Im} u \longrightarrow F^{\oplus m} \longrightarrow C \longrightarrow 0 .$$

Note that Supp $(K) \subsetneq Y$ and Supp $(C) \subsetneq Y$, since u is a generic isomorphism. But K and C are coherent on Y and hence we have $H^i(Y, K\otimes L^n) = H^i(Y, C\otimes L^n) = 0$ for all $i > 0$ and $n \gg 0$ (by the induction hypothesis). Now the cohomology exact sequences give that

$$\begin{array}{ccc} (H^i(Y,F\otimes L^n))^{\oplus m} & \xrightarrow{\sim} & H^i(Y, f_*(G)\otimes L^n) \\ & & \wr \| \\ & & H^i(X, G\otimes(f^*L)^n) \end{array}$$

and so $H^i(Y, F\otimes L^n) = 0$ for all $i > 0$ and $n \gg 0$. Hence L is ample on Y.

Proof of the lemma. Let $K(X)$ and $K(Y)$ be the fields of rational functions on X and Y respectively. Since f is finite and surjective, $K(X)$ is a finite algebraic extension of $K(Y)$. Then $[K(X) : K(Y)] = m =$ degree f.

Let U be an affine open set in X and A its coordinate ring. Since $K(X)$ is the quotient field of A, we can choose $s_1, \ldots s_m \in A$ such that $s_1, \ldots, s_m$ is a $K(Y)$-basis of $K(X)$. Let M be the subsheaf of $K(X)$ generated by the s_i's. Note that M is a coherent sheaf on X and $s_i \in H^o(X,M)$.

X and $s_i \in H^0(X,M)$.

Now consider the homomorphism $u\colon \mathcal{O}_Y^{\oplus m} \longrightarrow f_*(M)$ defined such that $e_i \longmapsto s_i$, where $\{e_i\}$ is the canonical basis of $\mathcal{O}_Y^{\oplus m}$. Clearly u is a generic isomorphism (by the choice of the s_i's).

Suppose a coherent sheaf F on Y is given. Then u induces a homomorphism of sheaves

$$\underline{\mathrm{Hom}}_{\mathcal{O}_Y}(f_*(M),F) \longrightarrow \underline{\mathrm{Hom}}_{\mathcal{O}_Y}(\mathcal{O}_Y^{\oplus m}, F) \xrightarrow{\sim} F^{\oplus m} .$$

But $\underline{\mathrm{Hom}}_{\mathcal{O}_Y}(f_*(M),F)$ has a structure of an $f_*(\mathcal{O}_X)$-module and f is finite. Hence we can write

$$\underline{\mathrm{Hom}}_{\mathcal{O}_Y}(f_*(M),F) = f_*(G)$$

for some coherent sheaf G on X. This completes the proof of the lemma.

<u>Proposition 4.6</u>. Let L be a line bundle on X. Suppose L is generated by its global sections and $L|_C$ is ample for every integral curve C in X. Then L is ample on X.

<u>Proof</u>. We can assume without loss of generality that X is integral. Since L is generated by its global sections, L defines a morphism $\varphi\colon X \longrightarrow \mathbb{P}_k^n$ for some $n > 0$. We have $L = \varphi^*(\mathcal{O}_{\mathbb{P}}(1))$.

We claim that φ is a finite morphism. For, otherwise some fibre of φ would contain a curve C in X. But then, $L|_C$ being ample, it follows by Exercise 3.2, that $\deg(L|_C) > 0$. This is a contradiction. Now by Proposition 4.4, we have $L = \varphi^*(\mathcal{O}_{\mathbb{P}}(1))$ is ample on X.

Exercise 4.7. Let $f: X \longrightarrow Y$ be a finite surjective morphism of complete non-singular schemes X and Y. Let $D = \sum n_i X_i$ be a divisor on X, where each X_i is a prime divisor on X. Let ν_i be the degree of the morphism $X_i \longrightarrow f(X_i)$. Let $f_*(D) = \sum n_i \nu_i f(X_i)$. Prove that $f_*(D)$ is ample if D is ample. Give an example to show that the converse is false.

Exercise 4.8. Let X, X' be complete schemes, and let D, D' be ample divisors on X and X', respectively. Show that $p_1^*D + p_2^*D'$ is ample on $X \times X'$, where p_1 and p_2 are the projections onto the two factors.

§5. Nakai's Criterion for Ampleness.

In this section we give Kleiman's proof of Nakai's theorem which gives a numerical criterion for a divisor on a complete scheme X to be ample.

We first recall some basic definitions connected with the Snapper-Kleiman intersection number symbol $(D_1.D_2,\ldots,D_t.Y)$ where $D_1,\ldots,D_t \in \text{Div}(X)$ and Y is a closed subscheme of X. Also we state (without proof) some results needed for our purpose. For details see Kleiman [2].

For a coherent sheaf G on X, we denote by $\chi(G)$ the Euler-Poincaré characteristic of G, namely $\sum_{i\geq 0}(-1)^i \dim_k (H^i(X,G))$. We have the

Theorem. (Kleiman [2], p. 295) Let F be a coherent sheaf on X and let $s = \dim(\text{Supp}(F))$. Let $L_1,\ldots,L_t$ be t line bundles on X. Then $\chi(L_1^{n_1} \otimes L_2^{n_2} \otimes \cdots \otimes L_t^{n_t} \otimes F)$ is a numerical polynomial in $n_1,\ldots,n_t$ of total degree $\leq s$.

By a numerical polynomial in $n_1,\ldots,n_t$, we mean a polynomial in $n_1,\ldots,n_t$ with rational coefficients which assumes integer values whenever $n_1,\ldots,n_t$ are integers.

Let $D_1,\ldots,D_t \in \text{Div}(X)$ and $L_i = \mathcal{O}_X(D_i)$. Let Y be a closed subscheme of X of dimension t. Then the intersection number of $D_1,\ldots,D_t$ with Y, denoted by $(D_1,\ldots,D_t.Y)$, is defined as the coefficient of the monomial $n_1\ldots n_t$ in $\chi(L_1^{n_1}\otimes\cdots\otimes L_t^{n_t}\otimes\mathcal{O}_Y)$.

We shall deviate a little in our notations from those of Kleiman. We write $(D_1\ldots D_t)$ for $(D_1\ldots D_t.X)$. Also we write $(D^t.Y)$ for

$(\underbrace{D \ldots D}_{t\text{-copies}}.Y)$.

Some basic results are

1) $(D_1 \ldots D_t.Y)$ is an integer, and is a symmetric t-linear form in $D_1,\ldots,D_t$.

2) $(D^t.Y)$ is easily seen to be $(t!)a$ where a is the coefficient of n^t in $\chi(L^n \otimes \mathcal{O}_Y)$ and $L = \mathcal{O}_X(D)$.

The following is the so-called Grothendieck-Kleiman-Moisezon-Mumford-Nakai-Zariski numerical criterion for ampleness (which we hereafter refer to simply as the Nakai criterion). (See Kleiman [1] and [2], Moisezon [1], and Nakai [2].)

<u>Theorem 5.1</u>. (<u>Nakai Criterion</u>) Let X be a complete scheme and D a divisor on X. Then D is ample $\Longleftrightarrow$ $(D^s.Y) > 0$ for every integral closed subscheme $Y \subseteq X$ of dimension s, for all $s \leq \dim(X)$.

<u>Proof</u>. Suppose D is ample. Replacing D by some nD, we can assume that D is very ample. Then X is a closed subscheme of $\mathbb{P} = \mathbb{P}_k^r$ for some $r > 0$. We have $D = X \cap H$ for some hyperplane $H \subseteq \mathbb{P}$. Now suppose $s \leq \dim X$ and Y an integral closed subscheme of X of dimension s. We have $(D^s.Y)_X = (H^s.Y)_{\mathbb{P}} = \deg(Y)$. But $\deg(Y) > 0$ since $Y \neq \emptyset$.

Conversely, suppose $(D^s.Y) > 0$ for every integral closed subscheme $Y \subseteq X$. Note that by Propositions 4.2 and 4.3, we can assume that X is integral. We prove the result by induction on $t = \dim X$. The result being trivial for $t = 0$, we can suppose $t \geq 1$.

Let $L = \mathcal{O}_X(D)$. By induction, we may assume $D|_Y$ is ample for all closed subschemes $Y < X$. We proceed in several steps.

(i) $\chi(L^n) \longrightarrow \infty$ as $n \longrightarrow \infty$. For, we have $(D^t) = (t!)a$ where a is the leading coefficient of (L^n). By hypothesis, we have $(t!)a > 0$ and so in particular $a > 0$.

(ii) We reduce to the case when D is effective.

For, we have L is a subsheaf of the sheaf of rational functions on X. Take $I = L^{-1} \cap \mathcal{O}_X$ and $J = I \otimes L$. Clearly I and J are non-zero coherent sheaves of ideals. Let Y and Z be the closed subschemes defined by I and J respectively. By induction we may assume that L_Y and L_Z are ample. So we have $H^i(Y,L_Y^n) = 0 = H^i(Z,L_Z^n)$ for all $i > 0$ and $n \gg 0$. But then the exact sequences

$$\begin{array}{ccccccccc} 0 & \longrightarrow & I\otimes L^n & \longrightarrow & L^n & \longrightarrow & L_Y^n & \longrightarrow & 0 \\ & & \| & & & & & & \\ 0 & \longrightarrow & J\otimes L^{n-1} & \longrightarrow & L^{n-1} & \longrightarrow & L_Z^{n-1} & \longrightarrow & 0 \end{array}$$

give that

$$H^i(X,L^n) \cong H^i(X,L^{n-1}) \quad \text{for } i \geq 2 \text{ and } n \gg 0.$$

Now from (i) it follows that $h^0(L^n)-h^1(L^n) \longrightarrow \infty$ as $n \longrightarrow \infty$ (where $h^i(L^n) = \dim_k (H^i(X,L^n))$), and so also $h^0(L^n) \longrightarrow \infty$ as $n \longrightarrow \infty$. Hence in particular $H^0(X,L^n) \neq (0)$. Thus replacing D by nD, we may assume D is effective.

(iii) We assert that L_n is generated by its global sections for $n \gg 0$.

For, consider the exact sequence

$$0 \longrightarrow L^{-1} \longrightarrow \mathcal{O}_X \longrightarrow \mathcal{O}_D \longrightarrow 0.$$

This gives the exact sequence

$$0 \longrightarrow L^{n-1} \longrightarrow L^n \longrightarrow L^n_D \longrightarrow 0 .$$

By induction we may assume L_D is ample. Hence $H^1(L^n_D) = 0$ for $n \gg 0$. Now the cohomology exact sequence

$$H^o(L^n) \longrightarrow H^o(L^n_D) \longrightarrow H^1(L^{n-1}) \longrightarrow H^1(L^n) \longrightarrow H^1(L^n_D)$$

gives that $H^1(L^{n-1}) \longrightarrow H^1(L^n)$ is surjective for $n \gg 0$. But the vector spaces $H^1(X,L^n)$ are all finite dimensional for $n \gg 0$. Hence the chain of dimensions

$$h^1(L^n) \geq h^1(L^{n+1}) \geq \cdots\cdots \geq \cdots\cdots$$

becomes stationary, i.e., $H^1(L^n) \xrightarrow{\sim} H^1(L^{n+1})$ for $n \gg 0$. Hence we get $H^o(L^n) \longrightarrow H^o(L^n_D)$ is surjective for $n \gg 0$. Since L_D is ample, by Theorem 3.1 (iii), L^n_D is generated by its global sections for $n \gg 0$. Hence by Nakayama, L^n is generated by its global sections for $n \gg 0$. This establishes (iii).

iv) To complete the proof of the theorem, we distinguish two cases:
(a) Suppose $\dim (X) = 1$. We have $(D) = \deg (D) > 0$. Hence by Exercise 3.2, D is ample.
(b) Suppose $\dim (X) \geq 2$. Since L^n is generated by its global sections, and since $L^n|_C$ is ample (by induction) for every integral curve C in X, we get by Proposition 4.6, that L^n is ample on X. Hence L is ample on X.

<u>Remark</u>. An example of D. Mumford (see §10, Appendix, below) shows that in the above criterion it is <u>not sufficient</u> to assume that $(D.C) > 0$ for every integral curve C in X in order that D be ample. In this example X is a non-singular surface but D is not effective. However, on a non-singular surface if D is effective and $(D.C) > 0$ for every integral curve C in X, then D is ample. Therefore one might ask: Suppose D is an <u>effective</u> divisor on X such that $(D.C) > 0$ for every integral curve C in X, then is D ample? The answer is NO, as shown by an example of C.P. Ramanujam. For details of these two examples, see §10 (Appendix), below.

§6. Pseudo-ampleness.

A divisor D on a complete scheme X is said to be pseudo-ample if $(D^s.Y) \geq 0$ for every closed integral subscheme $Y \subseteq X$ of dimension s, for all $s \leq \dim X$. This notion is due to Kleiman. In contrast to the Nakai criterion, we shall see that D is pseudo-ample if and only if $(D.Y) \geq 0$ for every integral closed curve Y in X. Using this we shall give a new criterion for ampleness, due to Seshadri, in terms of an intrinsic measure on the curves in X. Also we shall deduce another criterion for ampleness, due to Kleiman, in terms of a suitable norm on the space of curves in X.

A divisor D on X (complete as usual) is said to be numerically effective if $(D.Y) \geq 0$ for every closed integral curve Y in X. D is said to be numerically trivial, or numerically equivalent to zero, if $(D.Y) = 0$ for every closed integral curve Y in X.

Theorem 6.1. (Kleiman) Let X be a complete scheme and D a divisor on X. Then D is pseudo-ample $\Longleftrightarrow$ D is numerically effective.

Proof. Suppose D is numerically effective. We must prove that for all $s \leq \dim X$, $(D^s.Y) \geq 0$ for every integral closed subscheme $Y \subseteq X$ of dimension s. We may assume X is integral. We will use induction on s. The result is trivial for $s = 0$. We are reduced by induction to proving $(D^n) \geq 0$ where $n = \dim X$.

First observe that we can assume X is projective. For, by Chow's lemma, there is an integral projective scheme X' and a birational morphism $f: X' \longrightarrow X$. Let $D' = f^*D$. We have $\dim X' = n$, and $(D'^n) = (D^n)$ (projection formula for the intersection numbers: see Kleiman [2], Chap. I, §2, Prop. 6, p. 299). Thus replacing X by X', we may assume X is projective.

Fix a very ample divisor H on X. Let

$$m_i = (D^{n-i}.H^i) \qquad \text{for} \quad i = 0,1,\ldots,n.$$

Note that since H is ample, by the Nakai criterion, we have $m_n > 0$. Also by hypothesis and induction, we have $m_i \geq 0$ for $i = 1,\ldots,n-1$. We want to prove that $m_o \geq 0$.

Suppose to the contrary $m_o < 0$. Consider $F = aD + bH$ for positive integers a, b, and consider the polynomial

$$P(t) = m_o + \binom{n}{1}m_1 t + \ldots + m_n t^n .$$

We have

$$\begin{aligned}(F^n) &= a^n(D^n) + \binom{n}{1}a^{n-1}b(D^{n-1}.H)+\ldots+ b^n(H^n) \\ &= a^n m_o + \binom{n}{1}a^{n-1}b\, m_1 + \ldots + b^n m_n \\ &= a^n\, P(\tfrac{b}{a}) .\end{aligned}$$

Notice that $P(t)$ is an increasing function of t because $m_i \geq 0$ for $i = 1,\ldots,n$. Further we have $P(0) = m_o < 0$. Hence $P(t)$ has a single positive zero t_o, and $P(t) > 0$ for all $t > t_o$.

We claim that F is ample for all positive integers a,b such that $\frac{b}{a} > t_o$. For, we have

$$(F^s.Y) = a^s(D^s.Y) + \binom{s}{1}a^{s-1}b(D^{s-1}.H.Y) + \ldots + b^s(H^s.Y)$$

for every closed integral subscheme $Y \subseteq X$ of dimension s. For

$s < n$, we have by induction for D, $(D^{s-i}.H^i.Y) \geq 0$ for all $i = 0,\ldots,s-1$. Also since H is ample, we have $(H^s.Y) > 0$. Thus $(F^s.Y) > 0$. Moreover $(F^n) = a^n P(\frac{b}{a}) > 0$ by the choice of a,b. Hence by Nakai, F is ample. We may assume in fact F is very ample (replacing F by NF, i.e., b and a by Nb, Na respectively).

Note that $(D.F^{n-1}) \geq 0$ (because F is very ample $\Longrightarrow F^{n-1}$ is represented by an effective curve on X, so by hypothesis $(D.F^{n-1}) \geq 0$).

Consider the two polynomials

$$Q(t) = m_o + \binom{n-1}{1} m_1 t + \ldots + m_{n-1} t^{n-1}$$

$$R(t) = m_1 t + \binom{n-1}{1} m_2 t^2 + \ldots + m_n t^n .$$

We have $P(t) = Q(t) + R(t)$. Also for all positive integers a,b such that $\frac{b}{a} > t_o$, we have $(D.F^{n-1}) = a^{n-1} Q(\frac{b}{a}) \geq 0$. Therefore by continuity $Q(t) \geq 0$ for all $t \geq t_o$. Finally notice that $R(t) > 0$ for every $t > 0$ because $m_i \geq 0$ for $i = 1,\ldots,n-1$ and $m_n > 0$. Thus we get that

$$0 = P(t_o) = Q(t_o) + R(t_o) > 0$$

which is a contradiction. This completes the proof.

§7. Seshadri's Criterion for Ampleness.

Let X be a complete scheme. We denote by $m_P(C)$ the multiplicity of a point P on a curve C in X. For any curve C, let $m(C) = \sup_{P \in C} \{m_P(C)\}$. Note that $m(C) \geq 1$.

The following is a new criterion for ampleness which uses the measure $m(C)$ on the curves C in X.

Theorem 7.1. (Seshadri) Let X be a complete scheme and D a divisor on X. Then D is ample $\Longleftrightarrow \exists\, \epsilon > 0$ such that $(D.C) \geq \epsilon\, m(C)$ for every integral curve C in X.

Proof. Suppose D is ample. We can assume that D is very ample. So we can imbed X in a suitable projective space $\mathbb{P} = \mathbb{P}^n_k$ in such a way that $D = X \cap H$ for some hyperplane H in $\mathbb{P}$. Now for any integral curve C in X, we have $(D.C)_X = (H.C)_{\mathbb{P}} = \deg(C) \geq m(C)$. The result follows by taking $\epsilon = 1$.

Conversely, suppose $\epsilon > 0$ is such that $(D.C) \geq \epsilon\, m(C)$ for every integral curve C in X. Let $\dim(X) = n$. In view of Propositions 4.2 and 4.3, we can assume X is integral. Also by induction on $\dim X$, we can assume that $(D^s.Y) > 0$ for every closed subscheme $Y < X$ where $s = \dim Y$. By the Nakai criterion it suffices to prove that $(D^n) > 0$.

Take a non-singular closed point $P \in X$. Let X' be the blowing up of X at P. Let E be the exceptional divisor, and D' the total transform of D, i.e., if $f\colon X' \longrightarrow X$ is the canonical (birational) morphism, we have $E = f^{-1}(P) \approx \mathbb{P}^{n-1}_k$ and $D' = f^*(D)$.

Note that X' is a complete integral scheme of dimension n. Also we have $\mathcal{O}_X(E)|_E = \mathcal{O}_E(-1)$.

Now take an integer a such that $a > \frac{1}{\epsilon}$. Consider the divisor $G = aD' - E$. We assert that G is pseudo-ample on X'. For, by Theorem 6.1, it suffices to show that $(G.C') \geq 0$ for every integral curve C' in X'. We distinguish three cases:

i) $C' \subseteq E$. Then $(G.C') = a(D'.C') - (E.C')$

$$= 0 - (E|_E \cdot C')_E$$

$$= \deg_E C' > 0$$

(since $(E|_E.C')_E = (\mathcal{O}_E(-1).C')_E < 0$)

ii) $C' \cap E \neq \emptyset$ but $C' \nsubseteq E$. Write $C = f(C')$.

We have $(G.C') = a(D'.C') - (E.C')$

$$= a(D.C) - m_p(C) \quad .$$

But $a(D.C) \geq a\,\epsilon\, m(C) \geq m(C) \geq m_p(C)$, i.e., $(G.C') \geq 0$.

iii) $C' \cap E = \emptyset$. Now $(E.C') = 0$ and hence

$$(G.C') = a(D'.C') = a(D.C) \geq 0.$$

Now G being pseudo-ample, we have in particular $(G^n) \geq 0$. But

$$(G^n) = ((aD - E)^n) = a^n(D'^n) + (-1)^n(E^n)$$

because $(D'^{n-i}.E^i) = 0$ for $1 \leq i \leq n-1$.

Further $(D'^n)_{X'} = (D^n)_X$ since f is birational, and also

$(E^n)_{X'} = (E|_E^{n-1})_E = (-1)^{n-1}$. Thus we get

$$(G^n) = a^n(D^n) + (-1)^{2n-1}$$
$$= a^n(D^n) - 1 \geq 0$$

$\Longrightarrow (D^n) \geq \frac{1}{a^n} > 0$ which was to be proved.

This completes the proof of the theorem.

Corollary 7.2. Let X be a complete scheme and let D be a divisor on X. Then the property of being ample depends only on the numerical equivalence class of D.

Proof. Indeed, the integer $(D.C)$, where C is any curve on X, depends only on the numerical equivalence class of X.

§8. The Ample Cone.

Until the end of this chapter, we assume for simplicity that X is a non-singular complete scheme. (However, the results given are valid more generally, for example when X is quasi-divisorial in the sense of Kleiman [2], p. 326).

Let $F_1(X)$ denote the free abelian group generated by the set of all integral curves in X. We call any element $C \in F_1(X)$ also a curve (by abuse of language). If $C = \sum n_i C_i$, $n_i \geq 0$, we call C an effective curve.

We extend the intersection pairing to $F_1(X)$ by linearity: i.e., if $C = \sum n_i C_i$ and D is a divisor on X, then we define

$$(D.C) = \sum n_i (D.C_i) .$$

We say that two curves C_1 and C_2 are numerically equivalent if $(D.C_1) = (D.C_2)$ for all divisors D on X, and then write $C_1 \approx C_2$. We say that two divisors D_1 and D_2 are numerically equivalent (and write $D_1 \approx D_2$) if $D_1 - D_2$ is numerically equivalent to zero, i.e., $(D_1.C) = (D_2.C)$ for every curve C in X.

Now we define two real vector spaces:

$$A^1 = A^1(X) = (\mathrm{Div}\,(X)/\approx) \underset{\mathbb{Z}}{\otimes} \mathbb{R}$$

$$A_1 = A_1(X) = (F_1(X)/\approx) \underset{\mathbb{Z}}{\otimes} \mathbb{R} .$$

Clearly the intersection pairing $(D.C)$ induces a perfect pairing of A^1 with A_1. Here we state a theorem (without proof) needed in the sequel. For details see Kleiman [2], Chap. IV.

<u>Theorem</u>. The vector space A^1 is finite dimensional.

This is the Néron-Severi theorem (see Lang and Néron [1]) when X is non-singular and projective. Kleiman reduces the proof to the case when X is a projective non-singular surface.

The dimension of $A^1(X)$, denoted by $\rho = \rho(X)$, is called the <u>base number</u> or the <u>Picard number</u> of X. We note that ρ is also the dimension of $A_1(X)$.

We use D's (resp. C's) indiscriminately to denote divisors (resp. curves), their images in A^1 (resp. in A_1) and arbitrary elements of A^1 (resp. of A_1).

A subset K of a real vector space is called a <u>cone</u> if $K+K \subseteq K$ and $aK \subseteq K$ for every $a \in \mathbb{R}$, $a > 0$. It is clear that the closure $\bar{K}$ and the interior $\operatorname{int} K$ of a cone K are again cones.

The cone generated in A_1 by (the images of) the effective curves is called the <u>cone of curves</u> and is denoted by $A_1^+ = A_1^+(X)$. The closure $\bar{A}_1^+$ (of A_1^+ in A_1) is called the <u>closed cone of curves.</u> Note that the cone $P = P(X)$ generated by the pseudo-ample divisors in A^1, called the <u>pseudo-ample cone</u>, is just the closed dual of A_1^+, i.e.,

$$P = \{D \in A^1 \mid (D.C) \geq 0 \quad \text{for all } C \in A_1^+\} .$$

We define the <u>ample cone</u> $P^o = P^o(X)$ to be the cone in A^1 generated by the ample divisors on X if there are any, otherwise $P^o = \emptyset$.

We use the symbol $\| \ \|$ to denote any norm on the real vector space A_1. The following is another criterion for ampleness. It is an equivalent form of Kleiman's Theorem 2, p. 326 of [2], which says $P^{\circ} = \operatorname{int} P$. The proof is a corollary to Seshadri's criterion for ampleness.

Theorem 8.1. (Kleiman) Let X be a non-singular complete scheme and D a divisor on X. Then D is ample $\Longleftrightarrow \exists\, \epsilon > 0$ such that $(D.C) \geq \epsilon \|C\|$ for all integral curves C in X.

Proof. Suppose D is ample. We may assume D is very ample. So we can imbed X in a suitable projective space $\mathbb{P} = \mathbb{P}^n_k$ in such a way that $D = X \cap H$ for some hyperplane H in $\mathbb{P}$. Choose $D_1, \ldots, D_\rho$ a basis for A^1. We can assume that the D_i's and the D-D_i's are ample. The function $\| \ \|: A_1 \longrightarrow \mathbb{R}^+$ defined such that

$C \longmapsto \sum_{i=1}^{\rho} |(D_i.C)|$ is clearly a norm on A_1. Further we have

$\|C\| = \sum_i (D_i.C)$ for all integral curves C (since the D_i's are ample).

Hence we have $(D.C) - \|C\| = \sum_i ((D-D_i).C) > 0$ (since the D-D_i's are ample). This proves the result by taking $\epsilon = 1$.

Conversely, suppose $\epsilon > 0$ is such that $(D.C) \geq \epsilon \|C\|$ for all integral curves C in X. Then by Seshadri's criterion we have only to prove the

Lemma 8.2. Let X be a non-singular complete scheme. Let $\| \ \|$ be any norm on A_1. Then there exists a constant $\lambda > 0$ such that $m(C) \leq \lambda \|C\|$ for all integral curves C in X.

Proof. Let $\{U_i\}_{1 \leq i \leq m}$ be an affine open covering of X. Since X is non-singular, each $Y_i = X - U_i$ is an effective divisor on X. Let $D_i = nY_i$ and $L_i = \mathcal{O}_X(D_i)$. We can choose $n \gg 0$ such that for all i, $L_i|_{U_i}$ is generated by the global sections of L_i and these sections separate points of U_i.

Let C be an integral curve in X. Let $P \in C$. We have $P \in U_i$ for some i. We can find a $D_i' \in |D_i|$ such that $P \in D_i'$ and $C \not\subseteq D_i'$. But then we have $(D_i'.C) \geq m_P(C)$. Thus we get $m(C) \leq \sum_{i=1}^{m} (D_i.C)$.

Since the function $(D_i.C)$ is a linear functional on A_1 (for each i), we have $(D_i.C) \leq \lambda_i \|C\|$ for some $\lambda_i > 0$ and all C. Hence

$$m(C) \leq \sum_{i=1}^{m} (D_i.C)$$

$$\leq (\sum_{i=1}^{m} \lambda_i)\|C\| = \lambda\|C\|, \text{ say.}$$

This proves the lemma.

The following is a very useful corollary (and will be used in Kleiman's proof of Chevalley's conjecture).

Corollary 8.3. Let X be a non-singular complete scheme. Then X is projective $\Longleftrightarrow$ for all

$$C', C'' \in \overline{A_1^+}, \quad C' + C'' = 0 \Longrightarrow C' = C'' = 0.$$

Proof. Suppose X is projective. Choose a basis $D_1, \ldots, D_\rho$ of A^1 consisting of ample divisors. Let $C', C'' \in \overline{A_1^+}$ and suppose $C + C'' = 0$. We have $(D_i.(C'+C'')) = 0$ and $(D_i.C') \geq 0$, $(D_i.C'') \geq 0$ for all i. Hence $(D_i.C') = 0 = (D_i.C'')$ for all i. Hence $C' = C'' = 0$.

Conversely, the hypothesis is equivalent to saying that $\overline{A_1^+} \cap (-\overline{A_1^+}) = (0)$. Since X is non-singular, we have $A_1^+ \neq 0$. Using a standard result in the theory of cones, the hypothesis implies that there exists a divisor D such that $(D.C) > 0$ for every $C \in \overline{A_1^+}$ and $C \neq 0$. Hence we can find an $\epsilon > 0$ such that $(D.C) \geq \epsilon\|C\|$ for all $C \in \overline{A_1^+}$. In particular we have $(D.C) \geq \epsilon\|C\|$ for every integral curve C in X. Thus D is ample by the above theorem. Hence X is projective.

§9. Chevalley's Conjecture.

Theorem 9.1. (Chevalley's Conjecture) Let X be a non-singular compete scheme. Then X is projective $\Longleftrightarrow$ every finite set of closed points in X is contained in some open affine subset of X. (For a generalized form of this conjecture see Kleiman [2], Chapter IV, §2, Theorem 3, p. 327).

Proof. If X is projective, it is trivial that every finite set of closed points in X is contained in some open affine subset of X.

Conversely, in view of Corollary 8.3, to prove that X is projective we have only to prove that if $C', C'' \in \overline{A}_1^+$ and $C'+C'' = 0$ then $C' = C'' = 0$.

Let $C', C'' \in \overline{A}_1^+$. We have two sequences $\{C_n'\}$ and $\{C_n''\}$ in A_1^+ such that

$$C' = \varinjlim_n C_n' \quad \text{and} \quad C'' = \varinjlim_n C_n'' .$$

Suppose $C' + C'' = 0$. The proof that $C' = C'' = 0$ consists of 4 main steps.

1. We need the following two lemmas on finite-dimensional real vector spaces. The proofs are purely combinatorial and elementary (and are left as exercises).

Lemma 1. Let V be a real vector space of (finite) dimension n. Let S be a non-void subset of V, and let K be the cone generated by S. Then any element $x \in K$ can be written as $x = \sum_{i=1}^{n} \lambda_i s_i$ for some $s_i \in S$ and $\lambda_i \geq 0$.

<u>Lemma 2</u>. Let V be a finite-dimensional real vector space. Then given any basis $e_1,\ldots,e_n$ of V and any vector $\nu \in V$; $e_1 + m\nu,\ldots,e_n + m\nu$ is a basis of V for all $m \gg 0$.

Since the cone A_1^+ (in A_1 of dimension ρ) is generated by integral curves, by the lemma 1, we can write (for each n) that

$$C_n' = \sum_{i=1}^{\rho} a_n^{(i)} C_n^{(i)}$$

$$C_n'' = \sum_{i=\rho+1}^{2\rho} a_n^{(i)} C_n^{(i)}$$

where $a_n^{(i)} \geq 0$ and $C_n^{(i)}$ are integral curves for all $i = 1,\ldots,2\rho$.

2. We assert that there exists a subsequence $\{n_k\}$ and an open affine subset $U \subseteq X$ such that $C_{n_k}^{(i)} \nsubseteq X-U$ for all $i = 1,\ldots,2\rho$ and all n_k.

For, consider the pairs of the form $(\{n_k\},Z)$ where $\{n_k\}$ is a subsequence in $\mathbb{Z}^+$ and $Z \subseteq X$ is a closed subscheme satisfying

$$(*)\begin{cases} \text{(a) For every } n_k, \ \exists \text{ an } i \text{ such that } C_{n_k}^{(i)} \subseteq Z, \\ \text{(b) If } Y < Z \text{ is a proper closed subscheme of } Z, \\ \quad \text{there are only finitely many } (n_k,i) \text{ such that } \\ \quad C_{n_k}^{(i)} \subseteq Y. \end{cases}$$

First observe that such pairs do exist. For instance take a minimal closed subset $Z \subset X$ which contains infinitely many $C_n^{(i)}$. Then define a subsequence $\{n_k\}$ such that for each n_k, there is an i so that $C_{n_k}^{(i)} \subseteq Z$. Secondly note that such Z's are irreducible because in any case Z is a finite union of its irreducible components. Finally we claim that for a fixed subsequence $\{n_k\}$, if $Z_1,\dots,Z_s$ are distinct closed subsets of X satisfying (*), then $s \leq 2\rho$. To see this, we let $Y_p = \bigcup_{q \neq p} (Z_p \cap Z_q)$ where $p,q = 1,\dots,s$. We have $Y_p < Z_p$ by the irreducibility of Z_p, hence Y_p contains only finitely many $C_{n_k}^{(i)}$. So there exists an N such that for all $k \geq N$, and all $p = 1,\dots,s$ there is a $C^{(i_p)} \subseteq Z_p$ But $C_{n_k}^{(i_p)} \not\subseteq Y_p$. Hence $C_{n_k}^{(i_p)} \not\subseteq Z_q$ for any $q \neq p$. This shows that for an integer n_k $(k \geq N)$ there are s distinct integral curves $C_{n_k}^{(i_1)},\dots,C_{n_k}^{(i_s)}$ with $1 \leq i_1,\dots,i_s \leq 2\rho$. This is a contradiction unless $s \leq 2\rho$.

Now choose a subsequence $\{n_k\}$ and distinct closed subsets $Z_1,\dots,Z_s$ such that for each $p = 1,\dots,s$, $(\{n_k\},Z_p)$ satisfies (*) and s is maximum. Note that if W is a closed subset of X containing infinitely many $C_{n_k}^{(i)}$, then $Z_p \subseteq W$ for some p. Indeed, choose a minimal $W' \subseteq W$ containing infinitely many $C_{n_k}^{(i)}$, and then choose a subsequence $\{n_k'\}$ of $\{n_k\}$ such that $(\{n_k'\},W')$ satisfies (*). Now because every Z_p satisfies (*) for the subsequence $\{n_k'\}$ and s is maximum, we have $W' = Z_p$ for some p.

Choose closed points $z_p \in Z_p$ for each p. By hypothesis there exists an open affine subset U of X containing all z_p. Let $W = X - U$. Clearly $Z_p \not\subseteq W$ for every p. Hence W contains only finitely many $C_{n_k}^{(i)}$. By restricting the subsequence, we may assume that W contains no $C_{n_k}^{(i)}$. This proves the assertion 2.

3. We fix a subsequence $\{n_k\}$ and an affine open subset $U \subseteq X$ as in 2. Then we have $W = X - U$ is an effective divisor on X (since X is non-singular and U is affine open). Take divisors $D_1',\ldots,D_\rho'$ on X forming a basis for A^1. We claim that there is an integer $n > 0$ such that, if $D_j = D_j' + nW$, $j = 1,\ldots,\rho$, we have

(i) $D_1,\ldots,D_\rho$ is a basis for A^1 and

(ii) $(D_j.C) \geq 0$ for every integral curve C in X which meets U and all $j = 1,\ldots,\rho$.

To see this, let us write D' for any one of D_j', $j = 1,\ldots,\rho$; and first prove that there is an integer n' such that for all $n \geq n'$, we have $((D' + nW).C) \geq 0$ for every integral curve C in X which meets U. To do this, write $L' = \mathcal{O}_X(D')$ and take sections $g_1,\ldots,g_m$ of $L'|_U$ which generate $L'|_U$. Then for $n >> 0$, these sections extend to sections $f_1,\ldots,f_m$ of $\mathcal{O}_X(D'+nW)$ [EGA, I, 9.3.1]. But then if C is any integral curve in X meeting U, one of the f_i does not vanish identically on C. Hence $((D'+nW).C) \geq 0$ for all $n \geq n'$.

Now we can choose an integer n_o such that for all $j = 1,\ldots,\rho$ and all $n \geq n_o$, we have $((D_j'+nW).C) \geq 0$ for every integral curve C in X meeting U. By lemma 2, we may assume that the $D_j = D_j'+nW$ form a basis of A^1. This establishes 3.

4. To complete the proof of the theorem, note that by 2 and 3 (ii), we have

$$(D_j.C_{n_k}^{(i)}) \geq 0 \text{ for all } j = 1,\ldots,\rho;\text{ all } i = 1,\ldots,2\rho \text{ and all } n_k.$$

$\Longrightarrow (D_j.C'_{n_k}) \geq 0$ and $(D_j.C''_{n_k}) \geq 0$ for all j and all n_k

$\Longrightarrow (D_j.C') \geq 0$ and $(D_j.C'') \geq 0$ for all j (by continuity)

$\Longrightarrow (D_j.C') = 0 = (D_j.C'')$ for all j

(since $(D_j.C') + (D_j.C'') = 0$ by hypothesis)

$\Longrightarrow (D.C') = 0 = (D.C'')$ for all $D \in A'$ (by 3(i))

$\Longrightarrow C' = 0 = C''$.

§10. Appendix. Curves on Ruled Surfaces, and Examples of Mumford and Ramanujam.

Let C be a non-singular complete curve. Let E be a vector bundle of rank two on C. We consider the complete non-singular surface $X = \mathbb{P}(E) = \operatorname{Proj} \mathcal{S}$, when $\mathcal{S} = \sum_{n \geq 0} S^n(E)$. Then X is a so-called ruled surface. In fact, any minimal model of a surface birational to $C \times \mathbb{P}^1$, except for the projective plane, can be obtained in this way. (For a proof, see Hartshorne [5]; we will not use this fact.)

Let $\pi: X \longrightarrow C$ be the projection. Let $L = \mathcal{O}_X(1)$ be the canonical relatively very ample line bundle on X. Let D be the corresponding divisor. For any effective curve Y on X, we denote by $m(Y)$ the degree of Y over C. By linearity this gives a homomorphism $m: \operatorname{Pic} X \longrightarrow \mathbb{Z}$.

Proposition 10.1. There is an exact sequence

$$0 \longrightarrow \operatorname{Pic} C \xrightarrow{\pi^*} \operatorname{Pic} X \xrightarrow{m} \mathbb{Z} \longrightarrow 0 .$$

Proof. The proof is not difficult, and may be left to the reader. For hints, see Borel-Serre [1], §8, p. 115.

It follows that the divisors on X, modulo numerical equivalence, form a free abelian group of rank 2, generated by D and f, where f is any fibre of the projection π.

Now we will study effective curves Y on X by relating them to certain subsheaves of symmetric powers of E.

Proposition 10.2. For any $m > 0$ there is 1-1 correspondence between

a) effective curves Y on X (possibly reducible with multiple components), having no fibres as components, of degree m over C, and

b) sub-line bundles M of $S^m(E)$.

The correspondence is given by

$$Y \longmapsto \pi_* \mathcal{O}_X(m-Y) ,$$

where $\mathcal{O}_X(m-Y) = \mathcal{O}_X(m) \otimes \mathcal{O}_X(-Y)$, and

$$M \longmapsto \text{subschemes of } X \text{ defined by the homogeneous ideal } M\cdot\mathcal{S} \text{ generated by } M.$$

Furthermore, under this correspondence we have

$$(D.Y) = md - \deg M$$

where $d = \deg E$.

Proof. If Y is a curve of degree m, then $\mathcal{O}_X(m-Y)$ is trivial along the fibres, so $M = \pi_*\mathcal{O}_X(m-Y)$ is locally free of rank 1. It is a sub-line bundle of $\pi_*\mathcal{O}_X(m) = S^m(E)$ because Y contains no fibres as components. Conversely, it is clear that M defines a subscheme of degree m over C, and the two maps described are inverse to each other.

To calculate $(D.Y)$, we use the exact sequences on X,

$$0 \longrightarrow \mathcal{O}_X(m-Y) \longrightarrow \mathcal{O}_X(m) \longrightarrow \mathcal{O}_Y(m) \longrightarrow 0$$

$$0 \longrightarrow \mathcal{O}_X(m-1-Y) \longrightarrow \mathcal{O}_X(m-1) \longrightarrow \mathcal{O}_Y(m-1) \longrightarrow 0 .$$

Now since $\mathcal{O}_X(m-Y)$ is trivial along the fibres, $R^1\pi_*$ of it is a zero. Also $\mathcal{O}_X(m-1-Y)$ is like $\mathcal{O}_X(-1)$ along the fibres, so both

π_* and $R^1\pi_*$ are zero. Therefore we have exact sequences on C

$$0 \longrightarrow M \longrightarrow S^m(E) \longrightarrow \pi_*\mathcal{O}_Y(m) \longrightarrow 0$$

$$0 \longrightarrow S^{m-1}(E) \longrightarrow \pi_*\mathcal{O}_Y(m-1) \longrightarrow 0 .$$

Lemma 10.3. Let $\pi: Y \longrightarrow C$ be a finite morphism of curves, and let N be an invertible sheaf on Y. Then

$$\deg \pi_*N = \deg \pi_*\mathcal{O}_Y + \deg N.$$

Proof. Let L be an invertible sheaf on C. Then by the projection formula, $(\pi_*N)\otimes L = \pi_*(N\otimes\pi^*L)$. So
$\deg \pi_*(N\otimes\pi^*L) = \deg(\pi_*N)\otimes L = \deg \pi_*N + \deg \pi . \deg L.$
Also $\deg N\otimes\pi^*L = \deg N + \deg \pi . \deg L$. So the validity of the formula is not changed by replacing N by $N\otimes\pi^*L$. Now taking $\deg L \ll 0$, we may assume that N is a sheaf of ideals on Y, defining a subscheme Z. Then we have

$$0 \longrightarrow N \longrightarrow \mathcal{O}_Y \longrightarrow \mathcal{O}_Z \longrightarrow 0$$

$$0 \longrightarrow \pi_*N \longrightarrow \pi_*\mathcal{O}_Y \longrightarrow \pi_*\mathcal{O}_Z \longrightarrow 0$$

and so

$$\deg \pi_*N = \deg \pi_*\mathcal{O}_Y - \operatorname{length} \pi_*\mathcal{O}_Z .$$

But $\operatorname{length} \pi_*\mathcal{O}_Z = \operatorname{length} \mathcal{O}_Z = -\deg N$, so we are done.

Proof of Proposition, continued. From the two exact sequences above and the lemma, we obtain the equations

$$\deg S^m(E) = \deg M + \deg \pi_*\mathcal{O}_Y + m(D.Y)$$
$$\deg S^{m-1}(E) = \deg \pi_*\mathcal{O}_Y + (m-1)(D.Y) ,$$

because $\deg \mathcal{O}_Y(m) = m(D.Y)$. Now

$$\deg S^m(E) = \frac{1}{2}(m^2+m)d, \quad \text{where } d = \deg E$$

$$\deg S^{m-1}(E) = \frac{1}{2}(m^2-m)d.$$

Subtracting one equation from the other we have

$$md = \deg M + (D.Y)\ ,$$

as required.

Exercise 10.4. Show that $(D^2) = \deg E$.

The moral of this proposition is that to study intersection numbers of effective curves on X, one must know something about the degrees of sub-line bundles of $S^m(E)$.

In order to construct the example of Mumford, we first recall the definition of a stable bundle.

A vector bundle E on a curve C is said to be stable (resp. semi-stable) if for every sub-vector bundle $E' \subseteq E$, we have

$$(\deg E')/(\text{rank } E') < (\deg E)/(\text{rank } E)$$

(resp. $\leq$).

Theorem 10.5*. Let C be a curve of genus $g \geq 2$ over the complex numbers. Then

a) If E is stable, every symmetric power $S^m(E)$ is semi-stable, and

*This theorem was pointed out by C.S. Seshadri.

b) For any $r > 0$, $d \in \mathbb{Z}$, there exists a stable bundle E of rank r and degree d, such that all its symmetric powers $S^m(E)$ are stable. (In fact this holds for all "sufficiently general" stable bundles.)

Proof. We use the theorem of Narasimhan and Seshadri ([1], Theorem 2, p. 560] which says that stable bundles E correspond to irreducible unitary representations of a group π, where π is generated by elements $a_1, b_1, \ldots, a_g, b_g, c$ with relations

$$a_1 b_1 a_1^{-1} b_1^{-1} \cdots a_g b_g a_g^{-1} b_g^{-1} c = 1$$

$$c^N = 1$$

where N depends on the rank and degree of E.

Since the symmetric powers of a unitary representation are again unitary (possibly reducible), we see that if E is stable, then $S^m(E)$ is a direct sum of the stable bundles, hence semi-stable.

To prove b), we must construct an irreducible unitary representation of π all of whose symmetric powers are irreducible. For simplicity, we give the proof in the case of rank two (the proof in general being analogous).

Since $g \geq 2$, if we have chosen any two unitary matrices $A_1, B_1 \in U(2)$, then we can find further unitary matrices $A_2, B_2, \ldots, A_g, B_g, C$ satisfying the relations above. Thus it will be sufficient to find unitary matrices $A = A_1$ and $B = B_1$ such that for all $m > 0$, $S^m(A)$ and $S^m(B)$ form an irreducible pair, i.e., they have no common fixed subspace.

Let

$$A = \begin{pmatrix} \lambda_1 & 0 \\ 0 & \lambda_2 \end{pmatrix}$$

where $|\lambda_i| = 1$, and λ_2/λ_1 is not a root of unity. Then A is unitary. Furthermore, for any $m > 0$,

$$S^m(A) = \begin{pmatrix} \lambda_1^m & & 0 \\ & \lambda_1^{m-1}\lambda_2 & \\ 0 & & \lambda_2^m \end{pmatrix}$$

has all distinct eigenvalues, because of the choice of λ_1 and λ_2. Thus the only fixed subspaces of $S^m(A)$ are the subspaces spanned by some subset of the standard basis.

Consider an arbitrary unitary matrix

$$B = \begin{pmatrix} \mu_{11} & \mu_{12} \\ \mu_{21} & \mu_{22} \end{pmatrix}.$$

We claim that if B is sufficiently general, then all the entries of all the matrices $S^m(B)$ are non-zero. Indeed, for each i,j,m, the $(i,j)^{th}$ entry of $S^m(B)$ is given by a certain polynomial $\Phi_{i,j,m}(\mu_{11},\mu_{12},\mu_{21},\mu_{22})$, which is not identically zero. On the other hand, the set $U(2)$ of all unitary matrices is a real 4-manifold in $GL(2)$, which is a complex 4-manifold, and $U(2)$ is not contained in any analytic hypersurface of $GL(2)$. Hence by the Baire category theorem there exist unitary matrices $B = (\mu)$, such that none of the countably infinite set of polynomials $\Phi_{i,j,m}(\mu)$ is zero.

Now for any $m > 0$, $S^m(A)$ and $S^m(B)$ form an irreducible pair. Indeed, in order for $S^m(B)$ to have as fixed subspace a subspace generated by a subset of the standard basis, it would have to have some entries zero.

Example 10.6 (due to Mumford: see Kleiman [2], p. 326). There is a complete non-singular surface X, and a divisor D, such that $(D.Y) > 0$ for all effective curves Y, but D is not ample.

Let C be a non-singular curve of genus $g \geq 2$ over the complex numbers. Then by the above theorem, there exists a stable bundle E of rank two and degree zero, such that all its symmetric powers $S^m(E)$ are stable.

Let $X = \mathbb{P}(E)$, and let D be the divisor corresponding to $\mathcal{O}_X(1)$. If Y is a fibre of π, then $(D.Y) = 1$. If Y is an irreducible curve of degree m over C, then Y corresponds to a sub-line bundle $M \subseteq S^m(E)$. But $S^m(E)$ is stable of degree = 0, so $\deg M < 0$. Therefore

$$(D.Y) = md - \deg M > 0 \qquad (\text{since } d = 0) .$$

Thus $(D.Y) > 0$ for every effective curve Y on X.

On the other hand, D is not ample, because $(D^2) = d = 0$.

Problem 10.7. Does a similar example exist over a field k of characteristic $p > 0$? It would be sufficient to prove the existence of a curve with a stable bundle of rank two and degree zero, all of whose symmetric powers are stable.

<u>Example 10.8</u> (<u>Ramanujam</u>). There is a complete non-singular three-fold $\bar{X}$, and an <u>effective divisor</u> $\bar{D}$, such that $(\bar{D}.Y) > 0$ for all effective curves, but D is <u>not ample</u>.

Indeed, let X be a non-singular surface, and D a divisor with $(D.Y) > 0$ for all effective curves, and $(D^2) = 0$, as in the example of Mumford above. Let H be an effective ample divisor on X. Then we define $\bar{X} = \mathbb{P}(\mathcal{O}_X(D-H)\oplus\mathcal{O}_X)$, and let $\pi: \bar{X} \longrightarrow X$ be the projection. Then $\bar{X}$ is a compactification of the geometric vector bundle $\mathbb{V}(\mathcal{O}_X(D-H))$. Let $X_o \subseteq \bar{X}$ be the zero-section of this vector bundle, so that $(X_o^2) = (D-H)_{X_o}$. We define $\bar{D} = X_o + \pi^*H$.

Observe that $\bar{D}$ is effective by construction. Let Y be an irreducible curve on $\bar{X}$. To calculate $(\bar{D}.Y)$ we distinguish three cases.

1) If Y is a fibre of π, then

$$(\bar{D}.Y) = (X_o.Y) + (\pi^*H.Y) = 1 + 0 = 1.$$

2) If $Y \subseteq X_o$, then

$$\begin{aligned}(\bar{D}.Y) &= (\bar{D}|_{X_o}.Y)_{X_o}\\ &= (D-H+H.Y)_{X_o} = (D.Y)_{X_o} > 0.\end{aligned}$$

3) If $Y \not\subseteq X_o$, and $\pi(Y)$ is a curve Y' in X, then

$$(\bar{D}.Y) = (X_o.Y) + (\pi^*H.Y) .$$

But $(X_o.Y) \geq 0$, and $(\pi^*H.Y) = (H.Y')_X$ by the projection formula. But the latter is > 0 because H is ample.

Thus $(\overline{D}.Y) > 0$ for every effective curve on $\overline{X}$. On the other hand, $\overline{D}$ is not ample, because

$$(\overline{D}^2.X_o) = (\overline{D}\big|_{X_o}^2)_{X_o} = (D^2)_{X_o} = 0.$$

CHAPTER II

AFFINE OPEN SUBSETS

If Y is the support of an ample divisor on a complete scheme X, then one sees easily that the complement $U = X-Y$ is affine. In this chapter we ask to what extent is the converse true? Can one give necessary and sufficient conditions on Y, and on the linear systems of divisors with support on Y, for U to be affine? The question is not completely answered yet. However, we can give some necessary conditions and some sufficient conditions. And in the case of curves and surfaces, a complete solution is available. In the last section we give a necessary and sufficient condition for U to be affine, in terms of a suitable blowing-up of X with center contained in Y.

The main results of this chapter are due to Goodman [1]. We have followed his proofs quite closely. However, the criterion of Theorem 5.1 is new. Also, for convenience, we have included a first section giving Serre's cohomological characterization of affine schemes.

We now summarize the results of this chapter. If X is a complete non-singular variety of dimension 1 or 2, then U is affine $\Longleftrightarrow$ Y is the support of an ample divisor. Let X be a complete integral scheme of dimension ≥ 2, and let Y be a closed subset. Assume that U contains no complete curves. Then we have the following implications. (Here D will always denote a divisor with support Y.)

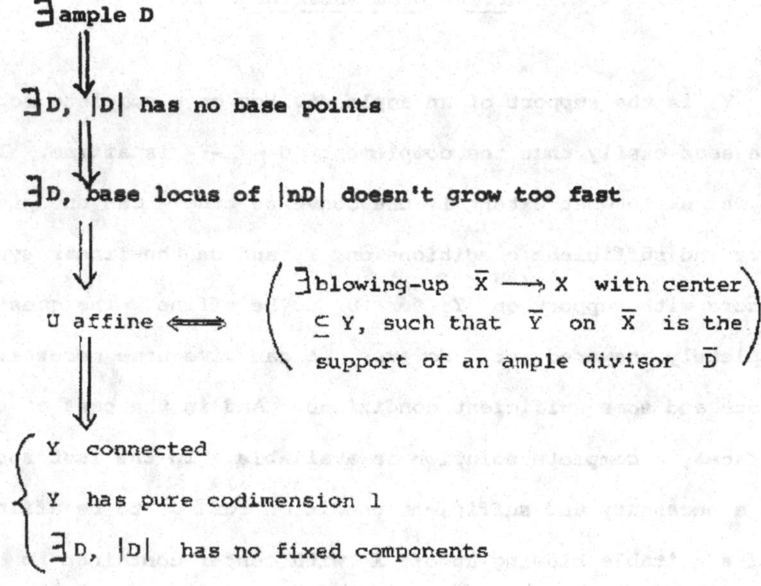
∃ ample D
∃ D, |D| has no base points
∃ D, base locus of |nD| doesn't grow too fast
U affine ⟺
∃ blowing-up X̄ ⟶ X with center ⊆ Y, such that Ȳ on X̄ is the support of an ample divisor D̄
Y connected
Y has pure codimension 1
∃ D, |D| has no fixed components

§1. Serre's Criterion for Affineness.

Theorem 1.1. (Serre - [EGA, II, 5.2.1, III, 1.3.1]):

Let S be a noetherian scheme. Then the following conditions are equivalent:

(i) S is an affine scheme,

(ii) For every coherent sheaf F on S, $H^i(S,F) = 0 \; \forall \, i \geq 1$,

(iii) For every coherent sheaf of ideals I, $H^1(S,I) = 0$.

Proof.

(i) $\Longrightarrow$ (ii) : [EGA, III, 1.3.1].

(ii) $\Longrightarrow$ (iii) : Obvious.

(iii) $\Longrightarrow$ (i) : [EGA, II, 5.2.1]. We shall prove this implication:

Let $A = H^0(S, \mathcal{O}_S)$. Let $\varphi = S \longrightarrow \mathrm{Spec}\, A$ be the canonical morphism. We will prove that φ is an isomorphism.

First we will show that there exist $f_1,\ldots,f_r \in A$ such that

1) the open sets $S_{f_i} = \{x \in S \mid f_i(x) \neq 0\}$ are affine and cover S, and

2) the ideal $(f_1,\ldots,f_r) = A$, i.e., $\{\mathrm{Spec}\,(A_{f_i})\}$ cover $\mathrm{Spec}\,(A)$.

Let $x \in S$ be a closed point, and let U be an affine open nbd of x. Let $Y = S-U$. Let I_Y (resp. $I_{Y\cup\{x\}}$) be the sheaf of ideals in $\mathcal{O}_S$ defining the closed subscheme Y (resp. $Y \cup \{x\}$). We have an exact sequence

$$0 \longrightarrow I_{Y\cup\{x\}} \longrightarrow I_Y \longrightarrow I_Y \otimes k(x) \longrightarrow 0$$

and hence the cohomology exact sequence

$$\ldots \longrightarrow H^{0}(S,I_{Y}) \longrightarrow H^{0}(S,I_{Y}\otimes k(x)) \longrightarrow H^{1}(S,I_{Y\cup\{x\}}) \longrightarrow \ldots .$$

By (iii), we have $H^{1}(S,I_{Y\cup\{x\}}) = 0$ and so the homomorphism $H^{0}(S,I_{Y}) \longrightarrow H^{0}(S,I_{Y}\otimes k(x))$ is surjective. Hence the section of $I_{Y}\otimes k(x)$ whose value at x is non-zero is the image of a section

$$f \in H^{0}(S,I_{Y}) \subseteq H^{0}(S,\mathcal{O}_{S}) = A.$$

But then $f(x) \neq 0$ and $f(y) = 0$ for $y \in Y$. Thus $S_f \subseteq U$. Now U being affine, it follows that S_f is affine. By quasi-compacity we can find $f_1,\ldots,f_r \in A$ such that $\{S_{f_i}\}$ is an affine open covering of S.

It then follows that ideal $(f_1,\ldots,f_r) = A$. Indeed, consider the homomorphism $\mathcal{O}_S^r \longrightarrow \mathcal{O}_S$ defined by the sections $f_1,\ldots,f_r$. This is surjective because for every $x \in S$, at least one of f_i is a unit in $\mathcal{O}_{S,x}$. This gives an exact sequence

$$0 \longrightarrow F \longrightarrow \mathcal{O}_S^r \longrightarrow \mathcal{O}_S \longrightarrow 0.$$

Consider the filtration for F

$$(0) \subseteq F \cap \mathcal{O}_S \subseteq F \cap \mathcal{O}_S^2 \subseteq \ldots \subseteq F .$$

For $k = 1,\ldots,r$, we have exact sequences:

$$0 \longrightarrow F \cap \mathcal{O}_S^{k-1} \longrightarrow F \cap \mathcal{O}_S^{k} \longrightarrow F \cap \mathcal{O}_S^{k}/F \cap \mathcal{O}_S^{k-1} \longrightarrow 0.$$

Note that each $F \cap \mathcal{O}_S^{k}/F \cap \mathcal{O}_S^{k-1}$ is isomorphic to an ideal in $\mathcal{O}_S$. By (iii) and induction on k, we get that $H^{1}(S,F\cap\mathcal{O}_S) = \ldots = H^{1}(S,F) = 0$.

Hence the homomorphism $H^o(S,\mathcal{O}_S^r) \longrightarrow H^o(S,\mathcal{O}_S)$ is surjective, i.e., the homomorphism $A^r \longrightarrow A$ defined by $f_1,\ldots,f_r$ is surjective. Hence $(f_1,\ldots,f_r) = A$.

Note that for any $f \in A$, we have $H^o(S_f,\mathcal{O}_S) = A_f$ [EGA, I, 9.33]. Now for each $i = 1,\ldots,r$, S_{f_i} being affine and $H^o(S_{f_i},\mathcal{O}_S) = A_{f_i}$, it is well known that $\varphi|_{S_{f_i}}$ is an isomorphism of S_{f_i} onto Spec (A_{f_i}). But then φ is an isomorphism because $\{S_{f_i}\}$ and $\{\text{Spec } (A_{f_i})\}$ are affine open coverings of S and Spec (A) respectively.

This completes the proof.

The proofs of the following corollaries are similar to those of Propositions 4.1, 4.2, 4.3, and 4.4, Chapter I.

Corollary 1.2. A closed subscheme of an affine scheme is affine.

Corollary 1.3. A noetherian scheme S is affine $\Longleftrightarrow$ S_{red} is affine.

Corollary 1.4. S is affine $\Longleftrightarrow$ Each irreducible component of S is affine.

Corollary 1.5. (Chevalley) Let $f: S \longrightarrow S'$ be a finite surjective morphism of noetherian schemes. Then S is affine $\Longleftrightarrow$ S' is affine.

§2. Sufficient Conditions for the Complement of a Subvariety to be Affine.

X = complete and integral scheme (over k)

Y = closed subscheme of X, and

U = X - Y.

Proposition 2.1. If there exists an effective ample divisor D on X such that Supp D = Y, then U is affine.

Proof. Replacing D by nD, we may assume that D is very ample. Then |D| defines a closed immersion $X \hookrightarrow \mathbb{P}^n$ for some $n > 0$ and $D = X \cap H$ for some hyperplane H in $\mathbb{P}^n$. By a suitable projective isomorphism of $\mathbb{P}^n$, we may assume that H is a coordinate plane in $\mathbb{P}^n$.

Now U = X-Y is a closed subscheme of the affine n-space $\mathbb{A}^n = \mathbb{P}^n - H$, and hence is affine.

Remark. It is natural to ask if the converse of the above proposition is true. The answer is NO, in general, because the existence of an ample divisor on X forces X to be projective. But there do exist non-projective complete varieties (see Nagata [2] and Hironaka [2]). However, we shall see later that the answer is YES in some cases, viz., when X is a curve (Proposition 4.1, below) or a non-singular surface (Theorem 4.2, below).

Proposition 2.2. Suppose there exists an effective divisor D on X such that Supp D = Y and |D| has no base points. Suppose further that U contains no complete curves. Then U is affine.

(The converse is false. See the remark in §4 below.)

Proof. Since $|D|$ has no base points, it defines a morphism $\varphi: X \longrightarrow \mathbb{P}^n$ for some $n > 0$. We have $D = \varphi^{-1}(H)$, for some hyperplane H in $\mathbb{P}^n$. Thus $U = \varphi^{-1}(\mathbb{P}^n - H) = \varphi^{-1}(\mathbb{A}^n)$. We claim that $\varphi: U \longrightarrow \mathbb{A}^n$ is a finite morphism. For, we have (i) φ is proper and (ii) φ is quasi-finite since U contains no complete curves. Hence φ is finite, in particular, affine. Hence U is affine.

Remark. The following theorem of Zariski shows that in the above proposition (when X is non-singular projective) it suffices to assume that D has only finitely many base points.

Theorem (Zariski [6], Theorem 6.2, p. 579). Let X be a non-singular projective variety. Suppose D is a divisor on X such that $|D|$ has only finitely many base points. Then $|nD|$ has no base points for $n \gg 0$.

§3. Necessary Conditions.

Proposition 3.1. If U is affine, then Y has pure codimension 1. (We will generalize this result in Chapter III, Corollary 3.6.)

Proof. Let $\overline{X}$ be the normalization of X and $f: \overline{X} \longrightarrow X$, the canonical morphism. Let $\overline{U} = f^{-1}(U)$ and $\overline{Y} = f^{-1}(Y)$. From the statement "$\overline{Y}$ has pure codimension 1" our conclusion follows by the finiteness of f. Hence we may assume that X is normal.

Let y be the generic point of an irreducible component of Y. Let $\mathcal{O}_y = \mathcal{O}_{X,y}$. The proposition is equivalent to proving that $\dim(\mathcal{O}_y) = 1$. Let $S_y = \operatorname{Spec}(\mathcal{O}_y)$ and $i: S_y \longrightarrow X$, the canonical (mono)morphism. Since X is a scheme and S_y is affine, it follows that i is an affine morphism [EGA, II, 1.2.3]. In particular we get that $S_y - \{y\} = i^{-1}(U)$ is affine, since U is affine.

Consider the local cohomology exact sequence:

$$0 \to H^0_y(S_y, \mathcal{O}_{S_y}) \longrightarrow H^0(S_y, \mathcal{O}_{S_y}) \longrightarrow H^0(S_y - \{y\}, \mathcal{O}_{S_y}) \longrightarrow H^1_y(S_y, \mathcal{O}_{S_y}) \longrightarrow 0 .$$

Now $S_y - \{y\}$ affine implies that $H^0(S_y, \mathcal{O}_{S_y}) \longrightarrow H^0(S_y - \{y\}, \mathcal{O}_{S_y})$ is not surjective. Therefore we have $H^1_y(S_y, \mathcal{O}_{S_y}) \neq 0$. Hence by Grothendieck [LC], Corollary 3.10, it follows that depth $(\mathcal{O}_y) \leq 1$. But then $\dim(\mathcal{O}_y) \leq 1$ (hence $\dim(\mathcal{O}_y) = 1$) because of the following:

Lemma (Serre [EGA, IV, 5.8.6]). Let A be normal domain. Then $\dim(A) \geq 2 \Longrightarrow \operatorname{depth}(A) \geq 2$.

We note the following result whose proof will be given later.

Proposition. Suppose $\dim X \geq 2$ and U is affine. Then Y is connected.

Proof. See Corollary 6.2 below (and also Corollary 3.9, Chapter III).

Exercise 3.2 (Goodman). Assume X is normal and U affine. Then there is a divisor D with support $D = Y$ such that $|D|$ has no fixed components.

§4. Curves and Surfaces.

Proposition 4.1. Every non-empty proper closed subset Y of a complete integral curve X is the support of an effective ample divisor. (Hence, in particular, every open $U \neq X$ is affine.)

Proof. Let $\overline{X}$ be the normalization of X and $f\colon \overline{X} \longrightarrow X$, the canonical (finite and surjective) morphism. Let Y be a non-empty proper closed subset of X. Let $\overline{Y} = f^{-1}(Y)$. By Exercise 4.7, Chapter I, it suffices to prove the result for the closed subset $\overline{Y}$ of $\overline{X}$. Thus we may assume that X is normal. Hence X is non-singular.

Then $D = \sum_{P_i \in Y} P_i$ is an effective Cartier Divisor on X. Since X is non-singular, it follows, by Exercise 3.2, Chapter I, that D is ample, since $\deg D > 0$. This completes the proof.

Note: Every complete curve is projective.

An example of Nagata [2] shows that if a complete surface has singularities, it need not be projective (even if the singularities are normal). Therefore it is not possible to characterize affine open subsets of such surfaces in terms of ample divisors. However, Zariski [4] has shown that a complete normal surface is projective if all its singularities lie in an affine open subset. More generally Goodman proves that a complete surface is projective if all its non-factorial points lie in an affine open subset. For simplicity, we will present his theorem only in the case of a non-singular surface (the proof is almost the same).

Theorem 4.2 (Goodman [1], Theorem 2, p. 168). Let X be a complete non-singular surface. Then the open subset U is affine $\Longleftrightarrow$ there exists an effective ample divisor D on X such that $\operatorname{Supp}(D) = Y = X - U$. (In particular, such a surface is projective (Zariski).)

Proof. In view of Proposition 2.1, we have only to prove $\Longrightarrow$: The proof consists of three main steps:

I. Suppose U is affine. Then we have the following:

(i) Y has pure codimension 1 (Proposition 3.1).

(ii) Y is connected (Corollary 6.2 below).

(iii) U contains no complete curve (clear).

(iv) If $Y_1,\dots,Y_t$ are the irreducible components of Y, there is a divisor $D' = \sum_{i=1}^{t} n_iY_i$ with all $n_i \geq 0$ such that $(D'.Y_i) \geq 0$ for all i and $(D'.Y_i) > 0$ for at least one i.

To see (iv), we choose a non-constant function $f \in \Gamma(U,\mathcal{O}_X)$. By adding an appropriate constant to f, if necessary, we may assume that for each i there is a $y_i \in Y_i$ such that $f(y_i) \neq 0$. Consider the two effective divisors $D = (f)_0$ and $D' = (f)_\infty$. Since $(f) = D-D'$, we have $D \sim D'$. Moreover, by construction, we have $\operatorname{Supp}(D') \subseteq Y$ and hence $D' = \sum_{i=1}^{t} n_iY_i$ for some n_i with $n_i \geq 0$. Now $(D.Y_i) \geq 0$ (since $Y_i \not\subseteq \operatorname{Supp}(D)$) for all i and $(D.Y_i) > 0$ for some i (since U contains no complete curve). Hence D', being linearly equivalent to D, has the properties $(D'.Y_i) \geq 0$ for all i and

$(D'.Y_i) > 0$ for some i. This proves (iv).

Since Y is connected, we can renumber the components Y_i, allowing repetitions if necessary, in such a way that $(Y_i, Y_{i+1}) > 0$ for all i. Also we may assume $(D'.Y_1) > 0$ by choosing Y_1 suitably. We will now let t denote the number of Y_i in the new list (with repetitions).

II. We show now, by induction, that for each $r = 1, \ldots, t+1$, there is a divisor $F_r = \sum n_i Y_i$ with $n_i \geq 0$ such that

(a) $(F_r . Y_i) \geq 0$ for all $i = 1, \ldots, t$,

(b) $(F_r . Y_i) > 0$ for all $i = 1, \ldots, \min(r, t)$,

(c) $Y_1, \ldots, Y_{r-1} \subseteq \text{Supp}(F_r)$.

Taking $F_1 = D'$, we find that the result is true for $r = 1$. Assume by induction that $F_1, \ldots, F_r$ have been constructed. We can assume $r \leq t$. Hence, by (b), we have $(F_r . Y_r) > 0$. Now choose an integer $k_r > 0$ such that $k_r (F_r . Y_r) + (Y_r^2) > 0$. We define $F_{r+1} = k_r F_r + Y_r$. Clearly $Y_1, \ldots, Y_r$ appear in F_{r+1} with positive coefficients and $\text{Supp}(F_{r+1}) = \text{Supp}(F_r) \cup Y_r$. We have $(F_{r+1} . Y_i) = k_r (F_r . Y_i) + (Y_r . Y_i)$. Hence by the choice of k_r, $(F_{r+1} . Y_r) > 0$. By induction and the ordering of the Y_i's, it follows that $(F_{r+1} . Y_i) \geq 0$ for all $i = 1, \ldots, t$. To see (b) we distinguish two cases:

1) Suppose $r < t$. Then $\min(r+1,t) = r+1$.

$$\text{Now } (F_{r+1}.Y_i) > 0 \begin{cases} \text{by induction if } i < r \text{ but } Y_i \neq Y_r \\ \text{by the choice of } k_r \text{ if } Y_i = Y_r \\ \text{by induction and the ordering if } i = r+1. \end{cases}$$

2) Suppose $r = t$. Then for $i = 1,\ldots,t$

$$(F_{t+1}.Y_i) > 0 \begin{cases} \text{by induction if } Y_i \neq Y_t \\ \text{by the choice of } k_t \text{ if } Y_i = Y_t. \end{cases}$$

III. The divisor F_{t+1} is ample.

Indeed, let C be any curve in X. Since U contains no complete curve, $C \not\subseteq U$. If $C \neq Y_i$ for any i, we have $(Y_i.C) \geq 0$ for all $i = 1,\ldots,t$ and $(Y_i.C) > 0$ for some i. Hence, by II (c), $(F_{t+1}.C) > 0$. On the other hand, if $C = Y_i$ for some i, $(F_{t+1}.C) > 0$ by II (b). Thus $(F_{t+1}.C) > 0$ for all curves C in X. But F_{t+1} is effective, and hence by Nakai's criterion F_{t+1} is ample.

Now the proof of theorem is complete because F_{t+1} is effective, ample and $\mathrm{Supp}\,(F_{t+1}) = Y$.

§5. Higher Dimensions.

On varieties of dimension ≥ 3, the question of characterizing open affine subsets becomes more difficult. There is an example of Zariski, reconstructed by Goodman [1], Chapter III, of a non-singular projective 3-fold X and an irreducible hypersurface Y, such that $U = X-Y$ is affine, but the linear system $|nY|$ has base points for every $n > 0$. This shows that Theorem 4.2 cannot be extended to higher dimensions, and that the converse of Proposition 2.2 is false.

However, it seems reasonable to expect that U will be affine if one can find divisors with support Y which do not have too many base points. In the example of Goodman and Zariski, there is a curve C which is the set of base points of $|nY|$ for all $n > 0$. Furthermore, C occurs with "multiplicity 1" in the sense that for each n, there is a divisor $D' \in |nY|$ which is non-singular along C.

We will prove one result of this sort, which says if the base locus of $|nD|$ doesn't grow too fast, then U is affine.

First we define the base locus as a scheme. Let D be an effective Cartier divisor on a complete integral scheme X, corresponding to an invertible sheaf L. Let L' be the subsheaf of L generated by $H^{o}(X,L)$. Then there is a unique sheaf of ideals I such that $L' = I \otimes L$. We define the <u>base locus</u> of $|D|$ to be the subscheme defined by I.

Theorem 5.1. Let X be a complete integral scheme, let Y be a closed subset, and let $U = X-Y$. Let D be an effective Cartier divisor with support Y. Assume that

(i) For each point $P \in Y$, there is an integer $n > 0$ and an irredundant primary decomposition

$$b_n = \mathfrak{q}_1 \cap \ldots \cap \mathfrak{q}_r$$

of the ideal b_n in $A = \mathcal{O}_{P,X}$ of the base locus B_n of $|nD|$, such that for each $i = 1,\ldots,r$,

$$\text{length}\, A_{\mathfrak{p}_i}/\mathfrak{q}_i A_{\mathfrak{p}_i} < n,$$

where $\mathfrak{p}_i$ is the prime ideal associated to $\mathfrak{q}_i$, and

(ii) U contains no complete curves.

Then U is affine.

Remarks. The hypothesis (i) will be satisfied in particular if B_n becomes constant for large n. That is what happens in the example of Goodman and Zariski: one can check that B_n is just the (reduced) curve C, for all n. Note also that (i) is trivial if P is not a base point of $|nD|$.

Proof of Theorem. Let $x \in A$ be a local equation for D. We will show that $x^{n-1} \in b_n$ for all $n \gg 0$. First note that $b_1 \supseteq b_2 \supseteq \ldots$, and $b_1 . b_n \subseteq b_{n+1}$ for all n. Furthermore $x \in b_1$, so in any case $x^n \in b_n$ for all n.

Now take the n of the condition (i). For each i, we work in the local ring $A_{\mathfrak{p}_i}$. By our hypothesis we have $\mathfrak{q}_i A_{\mathfrak{p}_i} \supseteq \mathfrak{p}_i^{n-1} A_{\mathfrak{p}_i}$. On the other hand, $x^n \in b_n \subseteq \mathfrak{p}_i$, so $x \in \mathfrak{p}_i$, so $x^{n-1} \in \mathfrak{p}_i^{n-1}$. Therefore

$$x^{n-1} \in \mathfrak{q}_i = A \cap \mathfrak{q}_i A_{\mathfrak{p}_i}.$$

This holds for all i, so $x^{n-1} \in b_n$. Being true for one particular n, it remains true for all larger n, since $x \in b_1$.

Note also that if $x^{n-1} \in b_n$ at a given point $P \in Y$, it is true also in a neighborhood. So by quasi-compacity we have $x^{n-1} \in b_n$ for all $P \in Y$ and all $n \gg 0$.

Fix such an n. Blow up the base locus B_n of $|nD|$. Let $f\colon \overline{X} \longrightarrow X$ be the blowing-up map. Let E be the exceptional divisor. Now I claim the linear system $|f^*(nD)-E|$ has no base points. Indeed, let L be the invertible sheaf corresponding to D. Let I be the sheaf of ideals of B_n. Then by definition, $I \otimes L^n$ is generated by global sections. Therefore $f^*(I\otimes L^n)$ is generated by global sections on $\overline{X}$. But $f^*(I)$ is the sheaf of ideals of the exceptional divisor E. So $f^*(I\otimes L^n)$ is the invertible sheaf corresponding to $f^*(nD) - E$. To say it is generated by global sections is to say the complete linear system $|f^*(nD) - E|$ has no base points.

Next we will show that $\operatorname{Supp}(f^*(nD) - E) = f^{-1}(Y)$. Indeed, since $x^{n-1} \in b_n$ at every point, the subscheme B_n is contained in

$(n-1)D$, and so $f^*((n-1)D)-E$ is non-negative divisor on $\overline{X}$. Therefore $\operatorname{Supp}(f^*(nD) - E) = \operatorname{Supp} f^*(nD) - E) = \operatorname{Supp} f^*D = f^{-1}Y$.

Now Proposition 2.2 applied to $f^*(nD) - E$ on $\overline{X}$ shows that $\overline{X} - f^{-1}(Y) = U$ is affine.

Another way of measuring the base points of a linear system is given by Goodman [1], p. 177. He calls a divisor D almost base-point free if for every point $P \in X$ and every $\epsilon > 0$, there exists an $n > 0$ and a divisor $D' \in |nD|$ such that $\operatorname{mult}_P D' < \epsilon n$. He then proves that an almost base-point free divisor is at least pseudoample (loc. cit., Corollary, p. 178).

This definition suggests the following conjecture.

Conjecture 5.2 (Goodman). Let X be a complete non-singular variety. Let Y be a closed subset of X, and let $U = X-Y$. Assume U contains no complete curves. Then U is affine $\Longleftrightarrow$ there exists an effective Cartier Divisor D with support Y which is almost base-point free.

§6. Goodman's Criterion.

Theorem 6.1 (Goodman [1], Theorem 1, p. 165). Let Y be a closed subset of a complete integral scheme X and let $U = X-Y$. Then U is affine $\Longleftrightarrow$ there exists a closed subscheme $Z \subseteq Y$ such that $\overline{Y} = f^{-1}(Y)$ is the support of an effective ample divisor on $\overline{X}$ where $f: \overline{X} \longrightarrow X$ is the birational blowing-up of X with center Z.

Proof. $\Leftarrow$). If $\overline{Y}$ is the support of an ample divisor, then $\overline{U} = \overline{X} - \overline{Y}$ is affine by Proposition 2.1. But $\overline{U}$ is isomorphic to U and hence U is affine.

$\Longrightarrow$). To prove the necessity of the condition, we use two results of Nagata [3] on blowing-up.

First we recall the definition of the join of two birational varieties: Let X_1 and X_2 be two birational schemes, say $f: U_1 \xrightarrow{\sim} U_2$ is an isomorphism where U_i are dense open subsets of X_i, $i = 1,2$. Identify U_1 and U_2 and write $U = U_1 \approx U_2$. Consider now the induced morphism $U \xrightarrow{\Delta} X_1 \times X_2$. Then the closed subscheme $\overline{\Delta(U)}$ in $X_1 \times X_2$ is called the join of X_1 and X_2 and is denoted by $J(X_1,X_2)$. Clearly we have a commutative diagram of birational morphisms:

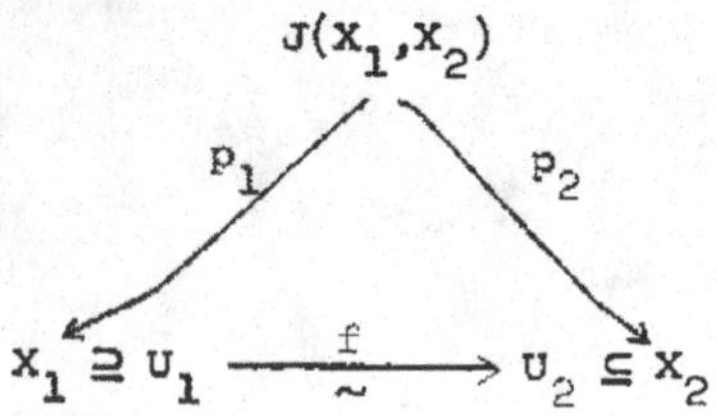

Note that when both X_1 and X_2 are complete (resp. projective) $J(X_1,X_2)$ is complete (resp. projective).

Theorem (Nagata [3], Theorem 3.2). Let X and X' be two birational varieties containing a dense open subset U. Then there exists a variety X^* dominating X and X' and containing U as a dense open subset. Moreover X^* can be obtained as the join of a finite number of varieties which are the blowings-up of X' with closed subschemes not meeting U as centers.

Theorem (Nagata [3], Theorem 3.3). Let X,X' and X^* be as above. If X is projective, then X^* can be obtained by blowing-up X' with respect to a single closed subscheme not meeting U.

Proof of Theorem, continued. Suppose U is affine. We have $U = \operatorname{Spec} A$ where $A = k[T_1,\dots,T_n]/J$. Hence U is closed in $\mathbb{A}^n$. Let $\mathbb{A}^n \longrightarrow \mathbb{P}^n$ be the canonical open immersion. Let X' be the closure of U in $\mathbb{P}^n$. Thus U is open and dense in X' and X' is projective. Obviously X and X' are birational. Moreover we have $Y' = X' - U = X' \cap H_\infty$ where H_∞ is the hyperplane $\mathbb{P}^n - \mathbb{A}^n$. Hence Y' is the support of a very ample divisor D' on X.

Now apply Nagata's first theorem to X and X'. Therefore there exists a variety X^* dominating X and X', and which is obtained as the join of the blowings-up of X' with respect to a finite number of closed subschemes, say $Z_1,\dots,Z_r \subseteq Y'$. Let $f\colon X^* \longrightarrow X$ be the dominant morphism.

We assert that there exists an effective ample divisor D^* on X^* with support equal to $Y^* = X^* - U$. For, let X_i be the blowing-up of X' with respect to the closed subscheme Z_i, $i = 1,\dots,r$. Consider, for example, the canonical morphism $f_1: X_1 \longrightarrow X'$. Note that f_1 is a projective morphism. If $E_1 = f_1^{-1}(Z_1)$ is the exceptional divisor on X_1, then $-E_1$ is a (relatively) very ample divisor on X_1. Hence by [EGA,II,4.6.13(ii)], $D_1 = nf_1^*(D') - E_1$ is ample on X_1 for all $n \gg 0$. Thus D_1 is an effective ample divisor on X_1 with support equal to $X_1 - U$. Thus we get D_i, effective ample divisors on X_i with $\mathrm{Supp}\,(D_i) = X_i - U$ for $i = 1,\dots,r$. But now $D = \sum_{i=1}^{r} p_i^*(D_i)$ is an ample divisor on $X_1 \times \dots \times X_r$ (Exercise 4.8, Chapter I). We have $X^* = J(X_1,\dots,X_r)$. It follows that $D^* = D|_{X^*}$ is ample and $\mathrm{Supp}\,(D^*) = X^* - U$. This proves the assertion. Note that since each X_i is projective, X^* is projective.

Now apply Nagata's second theorem to X^* and X. Hence we get a variety $\overline{X}$, dominating X^*, which is obtained as the blowing-up of X with respect to a single closed subscheme, say $Z \subseteq Y$. Let $g: \overline{X} \longrightarrow X^*$ be the dominant morphism, and hence we have a commutative diagram

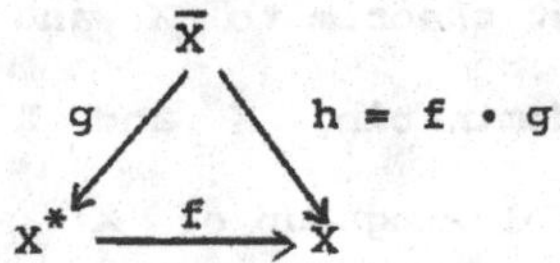

where $h: \overline{X} \longrightarrow X$ is the blowing-up morphism. If $\overline{E} = h^{-1}(Z)$ is the exceptional divisor on $\overline{X}$, then $-\overline{E}$ is relatively ample for h, hence also for g, so we get as before $n >> 0$, an ample divisor $\overline{D} = ng^*(D^*) - \overline{E}$ on $\overline{X}$. Now $\mathrm{Supp}\,(\overline{D}) = \overline{X} - U$, so we are done.

Corollary 6.2. Let X be a complete scheme with $\dim X \geq 2$. Suppose U is affine open in X. Then $Y = X - U$ is connected. (See Corollary 3.9, Chapter III for a generalization of this result.)

Proof. Let $h: \overline{X} \longrightarrow X$ be the birational blowing-up of X with respect to a closed subscheme $Z \subseteq Y$ as in the above theorem. Let $\overline{Y} = h^{-1}(Y)$. Since $h(\overline{Y}) = Y$, it suffices to prove that $\overline{Y}$ is connected. But $\overline{Y}$ is the support of an effective ample divisor on $\overline{X}$. Hence replacing X by $\overline{X}$, we may assume that X is projective and Y is the support of an effective ample divisor on X.

Let $\tilde{X}$ be the normalization of X, and let $f: \tilde{X} \longrightarrow X$ be the canonical finite surjective morphism. Let $\tilde{Y} = f^{-1}(Y)$. Hence by Proposition 4.4, Chapter I, $\tilde{Y}$ is the support of an effective ample divisor on $\tilde{X}$. But $\tilde{U} = \tilde{X} - \tilde{Y}$ is affine. Thus replacing X by $\tilde{X}$, we may assume that X is a normal (projective) variety.

We have $X \hookrightarrow \mathbb{P}^n$ and $Y = X \cap H$ for some hyperplane H in $\mathbb{P}^n$. Note that $\mathcal{O}_X(-H) = \mathcal{O}_X(-1)$ and $\mathrm{Supp}\,(\mathcal{O}_H) = Y$. For every integer $n > 0$, we have an exact sequence

$$0 \longrightarrow \mathcal{O}_X(-nH) \longrightarrow \mathcal{O}_X \longrightarrow \mathcal{O}_{nH} \longrightarrow 0 ,$$

and hence the cohomology exact sequence

$$0 \longrightarrow H^0(X,\mathcal{O}_X(-n)) \longrightarrow H^0(X,\mathcal{O}_X) \longrightarrow H^0(X,\mathcal{O}_{nH}) \longrightarrow H^1(X,\mathcal{O}_X(-n)).$$

Now look at the following

Lemma (Enriques-Severi-Zariski-Serre). Let X be a normal projective variety of dimension ≥ 2. Then for every line bundle L on X, $H^1(X,L(-n)) = 0$ for $n \gg 0$. In particular we have $H^1(X,\mathcal{O}_X(-n)) = 0$ for $n \gg 0$. For a proof see Serre [FAC], §76, Theorem 4, or Zariski [3].

This lemma implies therefore that $H^0(X,\mathcal{O}_X) \approx H^0(X,\mathcal{O}_{nH})$. But $H^0(X,\mathcal{O}_X) = k$ and hence $\dim_k(H^0(X,\mathcal{O}_{nH})) = 1$. This implies that the number of connected components of $\mathrm{Supp}\ (\mathcal{O}_{nH})(= Y)$ is precisely 1, i.e., Y is connected.

CHAPTER III

GENERALIZATION TO HIGHER CODIMENSIONS

In this chapter we begin our investigation of properties of a subvariety Y of a complete variety X, which are analogous to ampleness for divisors.

First we discuss ample vector bundles. It turns out that the notion of ample line bundle can be generalized quite satisfactorily to vector bundles of rank > 1. This has been done in Hartshorne [AVB], so we just recall the definition and basic results here, in §1.

In §2 we show that the tangent bundle to projective space is ample, and so the normal bundle to any non-singular subvariety of projective space is ample. This is very important for applications. In fact, it was one of the main motivations for developing the theory of ample vector bundles. As an amusing sidelight, we show that the projective plane can be characterized as the only complete non-singular surface with ample tangent bundle.

From some points of view, any subvariety of projective space can be considered to be "ample". So the ampleness of the normal bundle suggests that we should consider more generally any subvariety Y of a variety X whose normal bundle is ample. For example, any ample divisor has ample normal bundle. But the ampleness of the normal bundle is a purely local condition. Whereas the complement of an ample divisor is affine, we show here that the complement U of a divisor with ample normal bundle is a modification of an affine variety. This implies that all the cohomology groups $H^i(U,F)$ are finite-dimensional, for $i > 0$ and F coherent.

In generalizing, we can use the normal bundle to derive results about the formal scheme $\hat{X}$. In §3 we apply duality to relate the cohomology of $\hat{X}$ to the cohomology of $U = X-Y$. Thus we establish some relations between the dimension of Y and the cohomology of U. In §4 we consider subvarieties with ample normal bundle, and establish the finite-dimensionality of certain cohomology groups on U.

For the rest of the chapter, we are mostly concerned with subvarieties of projective space. In that case, we have fairly complete results relating properties of the subvariety $Y \subseteq \mathbb{P}^n$ to the cohomology of $U = \mathbb{P}^n - Y$. See §5 for a summary of these results.

The main sources for §§1-4 are Hartshorne [AVB] and [CDAV]. The results of §§6,7,8 are new.

§1. Ample Vector Bundles.

One way in which we can generalize the results of Chapter I is to replace line bundles by vector bundles of rank ≥ 1. This gives rise to the notion of an ample vector bundle, which we will discuss in this section. We will state the results without proof, because the techniques needed are very similar to those used for line bundles in Chapter I. For complete proofs, see Hartshorne [AVB].

We assume that X is a complete scheme over k.

By a vector bundle E on X, we mean a locally free sheaf (of $\mathcal{O}_X$-modules) E of finite rank.

We use the notation $\mathbb{V}(E)$ (resp. $\mathbb{P}(E)$) for the scheme vector bundle [EGA, II, 1.7.8] (resp. for the scheme projective bundle [EGA, II, 4.1.1]) associated to the vector bundle E on X. We note that the scheme $\mathbb{V}(E)$ (resp. $\mathbb{P}(E)$) is affine (resp. projective) over X. We denote by $\mathcal{O}_{\mathbb{P}(E)}(1)$ the tautological line bundle on $\mathbb{P}(E)$. It is clear that $\mathbb{P}(E)$ is a complete scheme over k. Also $\dim(\mathbb{P}(E)) = n+r-1$ if $\operatorname{rank} E = r$ and $\dim(X) = n$.

We say that a vector bundle E on a complete scheme X is ample if for every coherent sheaf F of $\mathcal{O}_X$-modules, $F \otimes S^n(E)$ is generated by its global sections for $n \gg 0$ (where $S^n(E)$ denotes the nth symmetric product of E).

Theorem 1.1. The following statements are equivalent for a vector bundle E on a complete scheme X.

(i) E is ample.

(ii) For every coherent sheaf F on X, we have

$H^i(X, F \otimes S^n(E)) = 0$ for all $i > 0$ and $n >> 0$.

(iii) The tautological line bundle $\mathcal{O}_{\mathbb{P}(E)}(1)$ on $\mathbb{P}(E)$ is ample.

Corollary 1.2. Suppose the rank of E is 1. Then the vector bundle E is ample $\Longleftrightarrow$ the line bundle E is ample.

Proposition 1.3. Let E be an ample vector bundle on X, and let Y be a closed subscheme of X. Then E restricted to Y is ample.

Proposition 1.4. Let E be a vector bundle on X. Then E is ample on X $\Longleftrightarrow E_{red}$ is ample on X_{red}.

Proposition 1.5. A vector bundle E on X is ample $\Longleftrightarrow E|_{X_i}$ is ample for every irreducible component X_i of X.

Proposition 1.6. Let $f: X \longrightarrow Y$ be a finite surjective morphism, and let E be a vector bundle on Y. Then E is ample on Y $\Longleftrightarrow f^*(E)$ is ample on X.

Proposition 1.7. Let $0 \longrightarrow E' \longrightarrow E \longrightarrow E'' \longrightarrow 0$ be an exact sequence of vector bundles on X. Then

i) E is ample $\Longrightarrow$ E" is ample, and

ii) E' and E" are ample $\Longrightarrow$ E is ample.

Corollary 1.8. Let $E_1, \ldots, E_n$ be vector bundles on X. Then $E = E_1 \oplus \ldots \oplus E_n$ is ample $\Longleftrightarrow$ each E_i is ample.

<u>Corollary 1.9</u>. Let E_1 and E_2 be two vector bundles on X. If E_1 is ample and E_2 is generated by its global sections, then $E_1 \otimes E_2$ is ample.

<u>Remark</u>. It is also true that <u>$E_1 \otimes E_2$ is ample whenever both E_1 and E_2 are ample</u>. This non-trivial result is proved by Hartshorne [AVB] when char k = 0, and by Barton [1] when char k $\neq$ 0. However, the above corollary gives this (in any characteristic) under the additional hypothesis that one of E_1 and E_2 is generated by its global sections.

<u>Proposition 1.10</u>. Suppose E is a vector bundle on X. Then E is ample $\Longleftrightarrow$ $S^n(E)$ is ample for $n >> 0$.

<u>Corollary 1.11</u>. X is projective $\Longleftrightarrow$ there exists an ample vector bundle on X.

<u>Proof</u>. Suppose E is an ample vector bundle on X. Then $S^n(E)$ is ample and generated by global sections for a suitable $n > 0$. If s is the rank of $S^n(E)$, then by Corollary 1.9, $(S^n(E))^{\otimes s}$ (and hence its quotient $\Lambda^s(S^n(E))$) is ample. But $\Lambda^s(S^n(E))$ is a line bundle which is ample. Hence X is projective.

<u>Note</u>. For further results about ample vector bundles, see Barton [1], Gieseker [1], Griffiths [1] and [2], Hartshorne [7], and Kleiman [5]. See also Chapter VI, §1 (i), below, for the analytic case.

§2. Ampleness of the Tangent Bundle on $\mathbb{P}^n_k$.

In this section we show that the tangent bundle to projective n-space is ample, and the normal bundle to any non-singular subvariety of projective space is ample. These results will be useful later (§§4 and 5, below) when we study subvarieties of projective space.

We also give a new characterization (Theorem 2.2, below) of the projective plane.

Let X be any scheme over k. We recall that the sheaf of 1-differentials $\Omega^1_{X/k}$ (of X/k) is (by definition) the sheaf of $\mathcal{O}_X$-modules I/I^2 where I is the sheaf of ideals in $\mathcal{O}_{X \times_k X}$ defining the closed subscheme X of $X \times_k X$ ([EGA, IV, 16.3.1]). Suppose $Y \subseteq X$ is a closed subscheme of X, say I_Y is the sheaf of ideals of $\mathcal{O}_X$ defining Y. Then we have an exact sequence of $\mathcal{O}_Y$-modules.

$$I_Y/I_Y^2 \longrightarrow \Omega^1_{X/k} \otimes \mathcal{O}_Y \longrightarrow \Omega^1_{Y/k} \longrightarrow 0 \quad ([\text{EGA, IV, } 16.4.21]).$$

Suppose now that Y is non-singular. Then the sequence

$$0 \longrightarrow I_Y/I_Y^2 \longrightarrow \Omega^1_{X/k} \otimes \mathcal{O}_Y \longrightarrow \Omega^1_{Y/k} \longrightarrow 0$$

is exact ([EGA, IV, 17.2.5]). Finally we have the

Theorem ([EGA, IV, 17.15.5, 17.15.6]). Let X be an integral scheme over k. Then X is non-singular $\Longleftrightarrow$ $\Omega^1_{X/k}$ is a locally free $\mathcal{O}_X$-module. Moreover when X is non-singular and $\dim(X) = n$, $\Omega^1_{X/k}$ is locally free of rank n.

For simplicity we assume that X is a complete non-singular variety. Suppose Y is a non-singular closed subscheme of X, and

I_Y is the ideal defining Y.

We recall that the tangent bundle T_X of X is the sheaf $\underline{\mathrm{Hom}}_{\mathcal{O}_X}(\Omega^1_{X/k}, \mathcal{O}_X)$, and the normal bundle $N_{Y/X}$ (of Y in X) is the sheaf $\underline{\mathrm{Hom}}_{\mathcal{O}_Y}(I_Y/I_Y^2, \mathcal{O}_Y)$.

The exact sequence

$$0 \longrightarrow I_Y/I_Y^2 \longrightarrow \Omega^1_{X/k} \otimes \mathcal{O}_Y \longrightarrow \Omega^1_{Y/k} \longrightarrow 0$$

gives the exact sequence

$$0 \longrightarrow T_Y \longrightarrow T_X \otimes \mathcal{O}_Y \longrightarrow N_{Y/X} \longrightarrow 0 .$$

By Propositions 1.3 and 1.7 (i), we get that $N_{Y/X}$ is ample on Y whenever T_X is ample on X.

Proposition 2.1. Let $\mathbb{P} = \mathbb{P}^n_k$. Then $T_{\mathbb{P}}$ is ample. Hence in particular $N_{Y/\mathbb{P}}$ is ample on Y for every non-singular subvariety $Y \subseteq \mathbb{P}$.

Proof. We have a canonical exact sequence

$$0 \longrightarrow \mathcal{O}_{\mathbb{P}} \longrightarrow (\mathcal{O}_{\mathbb{P}}(1))^{\oplus n+1} \longrightarrow T_{\mathbb{P}} \longrightarrow 0 .$$

Since $\mathcal{O}_{\mathbb{P}}(1)$ is ample, by Proposition 1.7, $T_{\mathbb{P}}$ is ample.

The projective space $\mathbb{P} = \mathbb{P}^n_k$, $n \leq 2$, is characterized by the ampleness of its tangent bundle $T_{\mathbb{P}}$. To be precise we have the

Theorem 2.2. For $n = 1,2$, the only complete non-singular variety X of dimension n whose tangent bundle is ample is the projective space $\mathbb{P}^n_k$.

Proof. For $n = 1$, X is a non-singular curve. The tangent bundle has degree 2-2g, where g is the genus of X. For it to be ample, we must have $2-2g > 0$, hence $g = 0$ and $X = \mathbb{P}^1_k$.

Now let X be a complete non-singular surface whose tangent bundle $T = T_X$ is ample. It follows that $\Lambda^2 T = -K$ is ample, where K is the canonical line bundle on X. Hence $H^0(X, \mathcal{O}_X(nK)) = 0$ for all $n > 0$. In particular, the geometric genus $p_g = \dim H^0(X, \mathcal{O}_X(K)) = \dim H^2(X, \mathcal{O}_X)$ and the second plurigenus $P_2 = \dim H^0(X, \mathcal{O}_X(2K))$ are zero.

Next we show that $H^0(X, \Omega^1_X) = 0$. If not, there would be a non-zero map of sheaves $\mathcal{O}_X \longrightarrow \Omega^1_X$. Hence there would be a non-zero map of the duals $\alpha: T \longrightarrow \mathcal{O}_X$. For any $n > 0$, this gives a non-zero map $S^n(\alpha): S^n(T) \longrightarrow \mathcal{O}_X$. Now T is ample. Hence for n sufficiently large, $\mathcal{O}_X(-1) \otimes S^n(T)$ is generated by global sections. But $\mathcal{O}_X(-1)$ has no global sections, and $S^n(\alpha)(-1): \mathcal{O}_X(-1) \otimes S^n(T) \longrightarrow \mathcal{O}_X(-1)$ is a non-zero map. This is impossible, so we conclude that $H^0(X, \Omega^1_X) = 0$.

Let $q = \dim \operatorname{Alb} X = \dim \operatorname{Pic}^0_{red} X$ be the irregularity of X. The natural map $X \longrightarrow \operatorname{Alb} X$ induces an injection $H^0(A, \Omega^1_A) \longrightarrow H^0(X, \Omega^1_X)$ where $A = \operatorname{Alb} X$. Hence by the previous step, we find $H^0(A, \Omega^1_A) = 0$ and so $A = 0$. Thus $q = 0$.

We have seen above that $p_g = H^2(\mathcal{O}_X) = 0$. Hence the Picard scheme $\operatorname{Pic}^0 X$ is reduced, and $q = \dim \operatorname{Pic}^0 X = \dim H^1(X, \mathcal{O}_X)$. Therefore $H^1(X, \mathcal{O}_X) = 0$. Finally, we conclude that the arithmetic genus $p_a = (\sum_i (-1)^i \dim H^i(X, \mathcal{O}_X)) - 1$ is zero.

Now by Zariski's proof of Castelnuovo's criterion "$p_a = P_2 = 0$", we find that X is a rational surface. We will use the classification of rational surfaces to complete the proof.

First note that since T is ample, for every non-singular curve $C \subseteq X$, its quotient $N_{C/X}$ is ample. Hence $(C^2) > 0$. It follows that X is a relatively minimal model. Furthermore, X cannot be a ruled surface over $\mathbb{P}^1_k$, because any fibre C would have $(C^2) = 0$. Thus X must be the projective plane $\mathbb{P}^2_k$.

References. Hartshorne [AVB, 2.6] for $\Lambda^2 T$ being ample; Chevalley [1] or Serre [7] for the Picard and Albanese varieties; Mumford [3], pp. 164, 198, for results on the Picard scheme; Zariski [4] for the proof of Castelnuovo's criterion; Hartshorne [5] or Nagata [3] for the classification of rational surfaces.

Problem 2.3. Does this result remain true for $n > 2$?

Exercise 2.4. If $Y \subseteq \mathbb{P}^n_k$ is a non-singular hypersurface of degree $d > 1$, show that its tangent bundle T_Y is *not* ample.

Hints. One method is as follows.

a) If T is ample, then $c_1(T)$ and $(c_1^2 - c_2)(T)$ are numerically positive, where c_i are the Chern classes of T (see Gieseker [1]). Deduce that $d < \frac{1}{2}n+1$.

b) If Y is a hypersurface of degree d in $\mathbb{P}^n_k$, with $d < n$, show that Y contains a straight line $Z \cong \mathbb{P}^1_k$.

c) If Z is a line in Y and T_Y is ample, then the normal bundle $N_{Z/Y}$ must be ample. Show that we must have deg $N \geq$ rank N, and deduce that $d \leq 1$.

§3. Cohomology of the Complement of a Subvariety.

Let S be a noetherian scheme of finite Krull dimension. In this section we define three integers $p(S)$, $q(S)$ and $cd(S)$ which depend on some cohomological properties of S. The main result of this section (Theorem 3.4, below) concerns a complete scheme X, a closed subset Y and the open set $U = X - Y$. We interpret the integers $p(U)$, $q(U)$ and $cd(U)$ in terms of the cohomology of certain sheaves on the formal completion $\hat{X}$ of X along Y. As corollaries we derive some general properties of these integers.

We define the cohomological dimension of S, written $cd(S)$, to be the smallest integer $n \geq 0$ such that $H^i(S,F) = 0$ for all $i > n$, and for all coherent sheaves F on S. Then by a well-known theorem of Grothendieck [1], Theorem 3.6.5, we have $cd(S) \leq \dim(S)$. By Serre's criterion for affineness (Theorem 1.1, Chapter II), we have S is affine if and only if $cd(S) = 0$. On the other hand we have Lichtenbaum's theorem (Corollary 3.5, below) which says (for an irreducible S) that S is proper if and only if $cd(S) = \dim(S)$.

Suppose now S is a (finite type) scheme over k. Then we define the integer $q(S)$ to be the smallest integer $n \geq -1$ such that $H^i(S,F)$ is a finite-dimensional k-vector space for all $i > n$, and for all coherent sheaves F on S. By the finiteness theorem of Serre-Grothendieck ([EGA, III, 3.2.1]), we have $q(S) = -1$ if S is proper over k.

Finally we define the integer $p(S)$ to be the largest integer n (or ∞) such that $H^i(S,F)$ is a finite-dimensional k-vector space for all $i < n$, and for all locally free sheaves F on S. We may remark that the integer $p(S)$ is of interest only when S is non-singular.

We have the following very useful

Proposition 3.1. Let S be a non-singular scheme (of finite type over k). Then the following conditions are equivalent (for any integer $r \geq 0$).

i) $cd(S)$ (resp. $q(S)$) $\leq r$,

ii) $H^i(S,F) = 0$ (resp. finite-dimensional over k) for all $i > r$, and all locally free sheaves F on S.

Further if S is also quasi-projective, then the above are equivalent to

iii) $H^i(S, \mathcal{O}_S(m)) = 0$ (resp. finite-dimensional over k) for all $i > r$, and all $m \ll 0$.

Proof. (i) $\Longrightarrow$ (ii): Obvious from definitions.

(ii) $\Longrightarrow$ (i):

We need the following

Lemma 3.2. (Kleiman). Let S be a non-singular scheme (of finite type over k). Then every coherent sheaf F on S is the quotient of a locally free sheaf on S.

(This result remains true more generally when S is divisorial. See Borelli [2], Theorem 3.3, p. 227).

Proof. Let $\{U_i\}_{1 \leq i \leq m}$ be an affine open covering of S. Since S is non-singular, each $D_i = S - U_i$ is the support of an effective

Cartier divisor. Let $L_i = \mathcal{O}_S(D_i)$, $1 \leq i \leq m$.

Suppose F is a coherent sheaf on S. Then by ([EGA, I, 9.3.1]), the global sections of $F \otimes L_i^n$ generate the stalks on U_i for $n \gg 0$ and each i. Thus we have integers n and m_i, $i = 1,\ldots,m$, and exact sequences

$$\mathcal{O}_S^{m_i} \longrightarrow F \otimes L_i^n \longrightarrow K_i \longrightarrow 0$$

with $\operatorname{Supp}(K_i) \subseteq D_i$. Hence the exact sequences

$$\mathcal{O}_S^{m_i} \otimes L_i^{-n} \longrightarrow F \longrightarrow K_i \otimes L_i^{-n} \longrightarrow 0 .$$

But $\bigcap_i D_i = \emptyset$, and so we obtain an exact sequence

$$\sum_i \mathcal{O}_S^{m_i} \otimes L_i^{-n} \longrightarrow F \longrightarrow 0 .$$

Since the $\mathcal{O}_S^{m_i} \otimes L_i^{-n}$ are locally free, the lemma is proved.

<u>Proof of</u> (ii) $\Longrightarrow$ (i):

We shall present a proof for the case $cd(S)$. It is similar for $q(S)$.

We must prove that $H^i(S,F) = 0$ for all $i > r$, and all coherent sheaves F on S. We do this by descending induction on i.

We have $H^i(S,R) = 0$ for all $i \gg 0$ (for example $i > \dim S$), and all coherent sheaves R on S. By induction, assume that $H^{i+1}(S,R) = 0$ for all coherent sheaves R on S.

Let F be a coherent sheaf on S. By the above lemma, we have an exact sequence

$$0 \longrightarrow R \longrightarrow E \longrightarrow F \longrightarrow 0$$

with E locally free, and R coherent on S. This gives the cohomology exact sequence

$$\cdots \longrightarrow H^i(S,E) \longrightarrow H^i(S,F) \longrightarrow H^{i+1}(S,R) \longrightarrow \cdots .$$

Now if $i > r$, $H^i(S,E) = 0$ (by (ii)), and $H^{i+1}(S,R) = 0$ by the induction hypothesis. Hence $H^i(S,F) = 0$, as required.

Suppose further S is also quasi-projective. Then (iii) $\Longrightarrow$ (i):

Let F be a coherent sheaf on S. We know that $F(m) = F \otimes \mathcal{O}_S(m)$ is generated by its global sections for $m \gg 0$. But then we have an exact sequence

$$0 \longrightarrow R \longrightarrow (\mathcal{O}_S(m))^{\oplus n} \longrightarrow F \longrightarrow 0$$

for a suitable integer $n > 0$, and $m \ll 0$ with R coherent on S. Now use the cohomology exact sequence, and proceed by induction as above. This completes the proof.

<u>Theorem 3.3</u>. (<u>Formal duality</u>) Let X be a complete non-singular variety of dimension n. Let $Y \subseteq X$ be a closed subset. Let $\wedge$ denote the formal completion along Y. Let $\omega = \Omega^n_{X/k}$ be the sheaf of n-differential forms on X, and let $G = \mathrm{Hom}_{\mathcal{O}_X}(F,\omega)$. Then

$$H^i(\hat{X},\hat{F}) \cong (H_Y^{n-i}(X,G))'$$

for any integer $i \geq 0$ where $'$ denotes dual vector space. (Note that these vector spaces need not be finite-dimensional.)

<u>Proof</u>. To prove this we need the following facts:

1. <u>Serre duality on</u> X (see Hartshorne [RD], Chapter VII, §4). Let X be a complete non-singular variety of dimension n, and let $\omega = \Omega^n_{X/k}$. Then

$$H^i(X,F) \cong (\mathrm{Ext}^{n-i}(F,\omega))'$$

for all $i \geq 0$, and for every coherent sheaf F on X. In particular, when F is locally free, we have

$$H^i(X,F) \cong H^{n-i}(X,G)'$$

for all $i \geq 0$.

2. Let X be a noetherian scheme, and let Y be a closed subset of X. Then

$$H^i_Y(X,F) \cong \varinjlim_m \mathrm{Ext}^i(\mathcal{O}_X/I^m_Y, F)$$

for every coherent sheaf F on X (see Grothendieck [LC], 2.8).

3. Let X be a scheme, and let Y be a closed subset of X. Then

$$H^i(\hat{X},\hat{F}) \cong \varprojlim_m H^i(Y_m,F_m)$$

for every coherent sheaf F on X, with Y_m being the subscheme of X defined by the sheaf of ideals I^m_Y, and $F_m = F \otimes \mathcal{O}_{Y_m}$ ([EGA, 0_{III}, 13.3.1]).

4. Let X be a noetherian scheme, and let M,N be coherent sheaves on X. Then

$$\mathrm{Ext}^i(F \otimes M,N) \cong \mathrm{Ext}^i(M, \underline{\mathrm{Hom}}(F,N))$$

for every locally free sheaf F on X (Exercise !).

<u>Proof of the theorem.</u> We have

$$H^i(\hat{X},\hat{F}) \quad = \quad \varprojlim_m (H^i(Y_m,F_m)) \qquad \text{(by (3))}.$$

But

$$\begin{aligned} H^i(Y_m,F_m) &= H^i(X,F_m) \\ &\cong (\mathrm{Ext}^{n-i}(F_m,\omega))' && \text{(by (1))} \\ &= (\mathrm{Ext}^{n-i}(\mathcal{O}_X/I_Y^m, G))' && \text{(by (4))} \end{aligned}$$

Hence

$$\begin{aligned} H^i(\hat{X},\hat{F}) &= \varprojlim_m (\mathrm{Ext}^{n-i}(\mathcal{O}_X/I_Y^m, G))' \\ &= (\varinjlim_m \mathrm{Ext}^{n-i}(\mathcal{O}_X/I_Y^m, G))' \\ &= (H_Y^{n-i}(X,G))' && \text{(by (2))}. \end{aligned}$$

<u>Theorem 3.4</u>. Let X be a complete non-singular variety of dimension n. Let $Y \subseteq X$ be a closed subscheme, and let $U = X - Y$. Let $r \geq 0$ be an integer. Then

(a) $q(U) \leq r \iff H^i(\hat{X},\hat{F})$ is finite-dimensional over k for all $i < n-r-1$, and all locally free sheaves F on X.

(b) $cd(U) \leq r \iff \alpha_i : H^i(X,F) \longrightarrow H^i(\hat{X},\hat{F})$ is an isomorphism for all $i < n-r-1$, and is injective for $i = n-r-1$, for all locally free sheaves F on X.

(c) $p(U) \geq r \iff H^i(\hat{X},\hat{F})$ is finite-dimensional over k for all $i > n-r-1$, and all locally free sheaves F on X.

Proof. Let F be a locally free sheaf on X. We use the long exact sequence of local cohomology

$$(*) \quad \cdots \longrightarrow H^i_Y(X,F) \xrightarrow{\beta_i} H^i(X,F) \longrightarrow H^i(U,F) \longrightarrow H^{i+1}_Y(X,F) \xrightarrow{\beta_{i+1}} \cdots$$

(See Grothendieck [LC, 1.9]).

Note that $H^i(X,F)$ is finite-dimensional for all i (since X is complete). So it follows that $H^i(U,F)$ is finite-dimensional if and only if $H^{i+1}_Y(X,F)$ is finite-dimensional.

Let $\omega = \Omega^n_{X/k}$ and let $G = \underline{\mathrm{Hom}}_{\mathcal{O}_X}(F,\omega)$. First observe that ω is a line bundle on X (since X is non-singular of dimension n, and hence $\Omega^1_{X/k}$ is locally free of rank n). So it follows that $\underline{\mathrm{Hom}}_{\mathcal{O}_X}(G,\omega) \cong F$. Thus we find that as F runs through the set of all locally free sheaves on X, so does G.

Now (a) and (c) are direct consequences of Proposition 3.1, Theorem 3.3, and the exact sequence (*) above.

To prove (b), consider the dual of the exact sequence (*) for the sheaf $G = \underline{\mathrm{Hom}}_{\mathcal{O}_X}(F,\omega)$:

$$(*)' \quad \cdots \longrightarrow H^{n-i+1}_Y(X,G)' \longrightarrow H^{n-i}(U,G)' \longrightarrow H^{n-i}(X,G)' \xrightarrow{\beta'_{n-i}} H^{n-i}_Y(X,G)' \longrightarrow \cdots$$

But

$$H^{n-i+1}_Y(X,G)' \cong H^{i-1}(\hat{X},\hat{F}) \qquad \text{(Theorem 3.3)}$$

and

$$H^{n-i}(X,G)' \cong H^i(X,F) \qquad \text{(Serre duality).}$$

Hence $(*)'$ can be written as

$$(*)' \cdots \xrightarrow{\alpha_{i-1}} H^{i-1}(\hat{X},\hat{F}) \longrightarrow H^{n-i}(U,G)' \longrightarrow H^i(X,F) \xrightarrow{\alpha_i} H^i(\hat{X},\hat{F}) \longrightarrow \cdots$$

Now note that $H^{n-1}(U,G) = 0$ if and only if α_{i-1} is surjective and α_i is injective. This establishes (b) (using Proposition 3.1).

This completes the proof of the theorem.

<u>Corollary 3.5</u>. (<u>Lichtenbaum's theorem for</u> $U = X-Y$). Let X be a complete non-singular variety of dimension n. Let $Y \subseteq X$ be a closed subset, and let $U = X-Y$. Then U is not complete $\Longleftrightarrow$ $cd(U) < \dim(U)$.

<u>Proof</u>. Take $r = n-1$. We get by Theorem 3.4 (b), that $cd(U) \leq n-1 \Longleftrightarrow H^o(X,F) \xrightarrow{\alpha_o} H^o(\hat{X},\hat{F})$ is injective for all locally free sheaves F on X. If Y is empty, then taking $F = \mathcal{O}_X$, we find that α_o cannot be injective (hence $cd(X) = n$). However, if Y is non-empty, then since F is locally free, every non-zero section of F has support on all of X, and so it gives a non-zero section of $\hat{F}$ over $\hat{X}$, i.e., α_o is injective.

<u>Remark</u>. This is a special case of the following result which has been proved in various degrees of generality by Grothendieck [LC], Theorem 6.9, Kleiman [3] and Hartshorne [CDAV], 3.2.

<u>Theorem</u> (<u>Lichtenbaum's theorem</u>). Let S be a scheme of finite type over k, of Krull dimension n. Then $cd(S) < n$ if and only if all irreducible components of S of dimension n are non-proper over k.

The following corollary generalizes the result that U affine $\Longrightarrow Y$ has pure dimension $n-1$ (Proposition 3.1, Chapter II).

Corollary 3.6. Let X be a non-singular projective variety, and let $Y \subseteq X$ be a closed subset, $U = X-Y$. Then $q(U) \leq r \Longrightarrow$ every irreducible component of Y has dimension $\geq n-r-1$.

Proof. By Theorem 3.4 (a), we have $q(U) \leq r \Longleftrightarrow H^i(\hat{X},\hat{F})$ is finite-dimensional for all $i < n-r-1$, and all locally free sheaves F on X.

Suppose the corollary is not true. Then Y has some irreducible component, say Y_o, of dimension s with $s < n-r-1$. Then we shall show that there exists an integer i (with $0 \leq i \leq s$) and a locally free sheaf F on X such that $H^i(\hat{X},\hat{F})$ is not finite-dimensional. This would contradict the hypothesis.

We do this in the following

Lemma 3.7. Let X be a projective scheme and $Y \subseteq X$ be a closed subset. Let

$$s = \min\left\{ \dim Y' \;\middle|\; \begin{array}{l} Y' \text{ is an irreducible component of } Y \\ \text{but not an irreducible component of } X \end{array} \right\}$$

Then there exists an integer i with $0 \leq i \leq s$ such that $H^i(\hat{X},\mathcal{O}_{\hat{X}}(-i))$ is an infinite-dimensional k-vector space.

Proof. We use induction on s. Suppose $s = 0$. Then there exists an isolated point y_o of Y which is not isolated in X. Therefore $\dim_k (\mathcal{O}_{X,y_o})$ is infinite. But we have

$$H^o(\hat{X},\mathcal{O}_{\hat{X}}) = H^o(Y-y_o,\ \mathcal{O}_{\hat{X}}) \oplus H^o(y_o,\ \hat{\mathcal{O}}_{X,y_o})$$
$$= H^o(Y-y_o,\ \mathcal{O}_{\hat{X}}) \oplus \hat{\mathcal{O}}_{X,y_o}$$

where $\hat{\mathcal{O}}_{X,y_o}$ is the usual completion of the local ring $\mathcal{O}_{X,y_o}$. Hence $H^o(\hat{X},\mathcal{O}_{\hat{X}})$ is infinite-dimensional.

Suppose $s \geq 1$. Let $Y_o \subseteq Y$ be an irreducible component of Y (but not of X) such that $\dim Y_o = s$. Choose a hyperplane $H \subseteq X$ such that $Y_o \nsubseteq H$. Then we have $H \cap Y_o \subseteq H \cap Y \subseteq H$ with $\dim (H \cap Y_o) = s-1$. Now by induction, there exists an integer i_o with $0 \leq i_o \leq s-1$ such that $H^{i_o}(\hat{H},\mathcal{O}_{\hat{H}}(-i_o))$ is infinite-dimensional.

Since H is a hyperplane we have an exact sequence

$$0 \longrightarrow \mathcal{O}_X(-1) \longrightarrow \mathcal{O}_X \longrightarrow \mathcal{O}_H \longrightarrow 0$$

and hence the exact sequence of formal completions

$$0 \longrightarrow \mathcal{O}_{\hat{X}}(-1) \longrightarrow \mathcal{O}_{\hat{X}} \longrightarrow \mathcal{O}_{\hat{H}} \longrightarrow 0 \quad .$$

Now twist this sequence by $\mathcal{O}_{\hat{X}}(-i_o)$ and write the cohomology exact sequence:

$$\cdots \longrightarrow H^{i_o}(\hat{X},\mathcal{O}_{\hat{X}}(-i_o)) \longrightarrow H^{i_o}(\hat{H},\mathcal{O}_{\hat{H}}(-i_o)) \longrightarrow H^{i_o+1}(\hat{X},\mathcal{O}_{\hat{X}}(-i_o-1)) \longrightarrow \cdots$$

Since the vector space in the middle is infinite-dimensional, we get that at least one of the extreme vector spaces is infinite-dimensional. Thus in any case we have an i with $0 \leq i \leq s$ such that $H^i(\hat{X},\mathcal{O}_{\hat{X}}(-i))$ is infinite-dimensional.

Corollary 3.8. Let X be a complete non-singular variety of dimension n. Let $Y \subseteq X$ be a closed subset, and let $U = X-Y$. Suppose every irreducible component of Y has dimension $\leq s$. Then

$$p(U) \geq n - s - 1 .$$

Proof. Since every irreducible component of Y has dimension $\leq s$, we have $\dim Y \leq s$. Since the underlying topological space of $\hat{X}$ is Y, by a theorem of Grothendieck [1], Theorem 3.6.5, we have

$$H^i(\hat{X},M) = 0$$

for all $i > s$, and all abelian sheaves M on $\hat{X}$. Hence in particular we have

$$H^i(\hat{X},\hat{F}) = 0$$

for all $i > s$, and all locally free sheaves F on X. Hence by Theorem 3.4 (c), we have $p(U) \geq n - s - 1$

Example. Suppose X is a non-singular complete scheme of dimension n. Let $Y \subseteq X$ be a finite subset, and let $U = X - Y$. Then $H^i(U,F)$ is finite-dimensional for all $i < n-1$, and all locally free sheaves F on X. Thus (using Corollary 3.6) we have $p(U) = q(U) = \mathrm{cd}\,(U) = n-1$.

The following corollary generalizes Proposition 6.2, Chapter II.

Corollary 3.9. Let X be a complete non-singular variety of dimension n. Let $Y \subseteq X$ be a closed subset, and let $U = X-Y$. Then $\mathrm{cd}\,(U) \leq n-2 \Longrightarrow Y$ is connected.

Proof. By Theorem 3.4 (b), we have

$\mathrm{cd}\,(U) \leq n-2 \Longrightarrow H^{o}(X,F) \xrightarrow{\alpha_o} H^{o}(\ddot{X},\ddot{F})$ is an isomorphism for all locally free sheaves F on X. Taking $F = \mathcal{O}_X$, we have in particular

$$k = H^{o}(X,\mathcal{O}_X) \xrightarrow{\cong} H^{o}(\hat{X},\mathcal{O}_{\hat{X}}) \ .$$

But $H^{o}(\hat{X},\mathcal{O}_{\hat{X}})$ contains one copy of k for each connected component of Y. Thus Y must be connected.

<u>Conjecture 3.10</u>. Let U be a complete non-singular irreducible scheme of dimension n. Then the following statements are equivalent:

(1) $p(U) \geq n-1$,

(2) there exists a complete irreducible scheme X containing U as an open subset such that $Y = X - U$ is the support of an effective divisor D on X with $\mathcal{O}_D(-D)$ ample on the subscheme D,

(3) there exists a complete irreducible algebraic space X containing U as an open subset such that $Y = X - U$ is a finite set of points.

<u>Remarks</u>. The implication (2)$\Longrightarrow$(1) is easy (see Exercise 4.14 below). The implication (2)$\Longrightarrow$(3) has been proved by Artin [5]. The implication (3)$\Longrightarrow$(2) should not be hard, once one has some familiarity with algebraic spaces (see Knutson [1] and Artin [2] and [5] for the theory of algebraic spaces). Thus the real problem is to show (1)$\Longrightarrow$(2). This is in some sense the "negative analogue" of Goodman's theorem (Theorem 6.1, Chapter II).

More generally, one can pose the <u>problem</u>: given a non-complete scheme U, describe in terms of cohomological properties of U how much one must add in order to obtain a complete scheme (or algebraic space) X.

EXERCISES AND PROBLEMS

<u>Exercise 3.11</u>. Let X and Y be quasi-projective schemes. Then show that $cd(X \times Y) = cd(X) + cd(Y)$. Does this remain true <u>without</u> the quasi-projectivity hypothesis?

<u>Exercise 3.12</u>. Let $f: X \longrightarrow Y$ be a smooth, proper morphism of relative dimension r. Then show that $cd(X) = cd(Y) + r$.

<u>Exercise 3.13</u>. Let X be a non-singular variety containing a non-singular subvariety V. Let $\tilde{X}$ be the <u>monoidal transformation</u> of X with center V. Then prove that

$$cd(\tilde{X}) = \max(cd(X), cd(V) + \operatorname{codim} V - 1).$$

<u>Problem 3.14</u>. Let $f: X \longrightarrow Y$ be a locally trivial algebraic fibre space with fibre Z. Is it true that $cd(X) = cd(Y) + cd(Z)$?

<u>Exercise 3.15</u>. Let S be a non-singular non-proper scheme of finite type over k. Show that there exists a locally free sheaf F and an integer i such that $H^i(S,F)$ is infinite-dimensional. Deduce that $p(S) \leq q(S)$.

§4. Subvarieties with Ample Normal Bundles.

Another way we can generalize the results of Chapter I is to consider a subvariety (of codimension ≥ 1) of a complete variety, whose normal bundle is ample. Of course this condition is only local around the subvariety, but still these subvarieties have many good properties analogous to those of ample divisors. In this section we give some first results, and state several conjectures and problems. For further properties of subvarieties with ample normal bundles, see §1 of Chapter V, below.

Let X be a complete scheme, and let $Y \subseteq X$ be a closed subscheme defined by the sheaf of ideals $I_Y \subseteq \mathcal{O}_X$. We say that Y is locally a complete intersection in X if the ideal sheaf I_Y of Y is locally generated by an $\mathcal{O}_X$-sequence. Note that I_Y is necessarily generated locally by $r = \text{codim}\,(Y,X)$ elements. (We recall that by a regular sequence (or an A-sequence) in a commutative ring A, we mean a sequence of elements $s_1,\ldots,s_r \in A$ such that s_1 is a non-zero divisor in A, and each s_i is a non-zero divisor in $A/(s_1,\ldots,s_{i-1})A$, for $i = 2,\ldots,r$).

For example any effective Cartier divisor Y on an integral scheme X is locally a complete intersection. If X and Y are both non-singular, then Y is locally a complete intersection (see Exercise 4.7, below).

<u>Theorem</u> (<u>Grothendieck</u> [EGA, IV, 16.9.4]). Let X be a complete scheme, and let $Y \subseteq X$ be a closed subscheme. Then the following are equivalent.

1) Y is locally a complete intersection in X,

2) (a) I_Y/I_Y^2 is a locally free $\mathcal{O}_Y$-module, and

(b) for each n, the canonical homomorphism

$$S^n_{\mathcal{O}_Y}(I_Y/I_Y^2) \longrightarrow I_Y^n/I_Y^{n+1}$$

is an isomorphism.

(The proof is not difficult; see Exercise 4.8, below).

Let X be a complete variety, and let $Y \subseteq X$ be locally a complete intersection. Then the locally free sheaf on Y

$$(I_Y/I_Y^2)^\vee = \underline{\mathrm{Hom}}_{\mathcal{O}_Y}(I_Y/I_Y^2, \mathcal{O}_Y)$$

is called the (<u>generalized</u>) <u>normal bundle of</u> Y <u>in</u> X, and is denoted by $N_{Y/X}$.

We remark that when X and Y are both non-singular, the generalized normal bundle of Y in X is simply the usual geometric normal bundle of Y in X. Now we give some examples of subvarieties Y of X whose normal bundles $N_{Y/X}$ are ample on Y.

<u>Examples</u>. 1. Let X be a complete variety, and let $Y \subseteq X$ be an effective ample Cartier divisor on X. Then we have $N_{Y/X} = \mathcal{O}_Y(Y)$, and so $N_{Y/X}$ is ample.

2. Let Y be a non-singular subvariety of the projective space $\mathbb{P} = \mathbb{P}^n_k$. Then by Proposition 2.1, $N_{Y/\mathbb{P}}$ is ample.

3. Let X be a projective variety, say X is closed in $\mathbb{P} = \mathbb{P}^n_k$. Suppose $Y \subseteq X$ is a closed subscheme, and is a proper intersection in $\mathbb{P}$, i.e., there exists a closed subscheme $Z \subseteq \mathbb{P}$ such that $Y = Z \cap X$ as subschemes in $\mathbb{P}$, with $\dim Y = (\dim X + \dim Z) - n$. It is easy to see that $N_{Y/X} \cong N_{Z/\mathbb{P}}|_Y$. Hence if $N_{Z/\mathbb{P}}$ is ample on Z (for instance when Z is non-singular), $N_{Y/X}$ is ample on Y.

4. Let X be a complete variety, and let $Y \subseteq X$ be a closed subscheme of codimension r. Suppose $Y = H_1 \cap \ldots \cap H_r$ (as subschemes in X) where the H_i are ample effective Cartier divisors on X. Then $N_{Y/X} \cong \bigoplus_i N_{H_i/X}|_Y$, and so $N_{Y/X}$ is ample.

5. Let X be a complete variety, and let $Z \subseteq Y \subseteq X$ be closed subschemes. Suppose that Z is locally a complete intersection in Y, and Y is locally a complete intersection in X. If $N_{Z/Y}$ and $N_{Y/X}$ are ample, then $N_{Z/X}$ is ample. Indeed, we have an exact sequence

$$0 \longrightarrow N_{Z/Y} \longrightarrow N_{Z/X} \longrightarrow N_{Y/X}|_Z \longrightarrow 0. \quad \text{[EGA, IV, 19.1.5 (iii)]}.$$

Now the assertion is a consequence of Propositions 1.3 and 1.7.

6. Let $f\colon X \longrightarrow X'$ be a finite surjective morphism of the complete varieties X and X'. Let $Y' \subseteq X'$ be locally a complete intersection, and let $Y = f^{-1}(Y')$. Then $N_{Y/X}$ is ample if and only if $N_{Y'/X'}$ is ample. Indeed, we have $N_{Y/X} = f^*(N_{Y'/X'})$. So the assertion follows by Proposition 1.6.

Let X be a complete scheme. For any vector bundle E on X, we denote by $\Gamma^n(E)$ the sheaf $S^n(\check{E})^\vee$ where $^\vee$ denotes the dual.

We say that a vector bundle E on X is <u>Γ-ample</u> if for every coherent sheaf F on X,

$$H^i(X,F\otimes\Gamma^n(E)) = 0$$

for every $i > 0$, and all $n >> 0$.

Note that if char $k = 0$, we have $S^n(E) \cong \Gamma^n(E)$ (Exercise 4.9, below), and hence a vector bundle E on X is Γ-ample if and only if E is ample. However, if char $k \neq 0$, an example of Gieseker (Exercise 4.10, below) shows that ampleness and Γ-ampleness are not equivalent. If rank $E = 1$, then $S^n(E) = \Gamma^n(E) = E^{\otimes n}$, so in that case ampleness and Γ-ampleness are equivalent. Finally we remark that ampleness and Γ-ampleness are equivalent if X is a non-singular curve over k of any characteristic. (Exercise 4.11, below).

<u>Theorem 4.1</u>. Let X be a complete non-singular variety of dimension n and let $Y \subseteq X$ be a non-singular closed subvariety of dimension s. Suppose that $N_{Y/X}$ is Γ-ample. Then

(a) $H^i(\hat{X},\hat{F})$ is a finite-dimensional k-vector space for all locally free sheaves F on X, and all $i < s$, and

(b) $p(U) = q(U) = n-s-1$, where $U = X-Y$.

<u>Proof</u>. (a) Let F be a locally free sheaf on X. We have $H^i(\hat{X},\hat{F}) = \varprojlim_m H^i(Y_m,F_m)$ where Y_m is the subscheme defined by I_Y^m and $F_m = F \otimes_{\mathcal{O}_X} \mathcal{O}_{Y_m}$ ([EGA, 0_{III}, 13.3.1]).

By Serre-Grothendieck [EGA, III, 3.2.1], we know that the vector spaces $H^i(Y_m, F_m)$ are finite-dimensional for all i. So to prove (a), clearly it suffices to show that the canonical maps

$$\alpha_{m+1,m} : H^i(Y_{m+1}, F_{m+1}) \longrightarrow H^i(Y_m, F_m)$$

are isomorphisms for all $i < s$, and $m \gg 0$.

The exact sequence

$$0 \longrightarrow I_Y^m/I_Y^{m+1} \longrightarrow \mathcal{O}_X/I_Y^{m+1} \longrightarrow \mathcal{O}_X/I_Y^m \longrightarrow 0$$

gives the cohomology exact sequence

$$\cdots \longrightarrow H^i(Y, F_1 \otimes (I_Y^m/I_Y^{m+1})) \longrightarrow H^i(Y, F_{m+1}) \xrightarrow{\alpha_{m+1,m}} H^i(Y, F_m) \longrightarrow \cdots$$

Thus to prove that $\alpha_{m+1,m}$ is an isomorphism for all $m \gg 0$, it suffices to show that $\alpha_{m+1,m}$ is injective for all $m \gg 0$ (by the finite-dimensionality). In other words, it suffices to show that

$$H^i(Y, F_1 \otimes (I_Y^m/I_Y^{m+1})) = 0$$

for all $i < s$, and all $m \gg 0$. To show this, we use Serre duality on Y (see Hartshorne [RD, Chapter VII, §4).

Let $\omega_Y = \Omega^s_{Y/k}$. Then by duality on Y, we have

$$H^i(Y, F_1 \otimes (I_Y^m/I_Y^{m+1})) \cong \operatorname{Ext}^{s-i}_{\mathcal{O}_Y}(F_1 \otimes (I_Y^m/I_Y^{m+1}), \omega_Y)'$$

where $'$ denotes the dual vector space. Since I_Y^m/I_Y^{m+1} is locally free, we have

$$\operatorname{Ext}^j_{\mathcal{O}_Y}(F_1 \otimes (I_Y^m/I_Y^{m+1}), \omega_Y) \cong H^j(Y, F_1^\vee \otimes \omega_Y \otimes (I_Y^m/I_Y^{m+1})^\vee) .$$

But $(I_Y^m/I_Y^{m+1})^\vee = (S^m(I_Y/I_Y^2))^\vee$

$$= (S^m(N_{Y/X}^\vee))^\vee$$

$$= \Gamma^m(N_{Y/X}) .$$

Thus we get that

$$H^i(Y,F_1\otimes(I_Y^m/I_Y^{m+1})) \cong H^{s-i}(Y,F_1^\vee\otimes\omega_Y\otimes\Gamma^m(N_{Y/X}))' .$$

Now the result (a) follows at once because $N_{Y/X}$ is Γ-ample, and $F_1^\vee\otimes\omega_Y$ is coherent on Y.

(b) By Theorem 3.4 (a), (a) is equivalent to $q(U) \leq n-s-1$. On the other hand if $q(U) < n-s-1$, by Corollary 3.6, we get that $\dim Y \geq s+1$. This is not possible. Hence $q(U) = n-s-1$.

By Corollary 3.8, we have $p(U) \geq n-s-1$. Suppose if possible $p(U) \geq n-s$. Then by Theorem 3.4 (c), $H^i(\hat{X},\hat{F})$ is finite-dimensional for all $i > s-1$, and all locally free sheaves F on X. But then in view of (a), $H^i(\hat{X},\hat{F})$ is finite-dimensional for all i, and all locally free sheaves F on X. Hence by Theorem 3.4 (a), we get (for example) that $q(U) \leq n-s-2$. This is a contradiction. Hence $p(U) = n-s-1$.

This completes the proof.

Remark. The above result holds also under the weaker hypothesis that Y is locally a complete intersection and $N_{Y/X}$ is Γ-ample. The proof is the same, except that in the duality theorem, one must take ω_Y to be the sheaf of dualizing differentials on Y. For details see Hartshorne [RD], Chapter V.

Theorem 4.2. Let X be a complete variety, and let $Y \subseteq X$ be an effective Cartier divisor. Then the following are equivalent.

(i) $N_{Y/X}$ is ample,

(ii) There exists a birational morphism $f: X \longrightarrow X'$ (of complete varieties) such that f is an isomorphism in a neighborhood of Y, and $Y' = f(Y)$ is an ample effective divisor on X'.

Proof. It is clear that (ii) $\Longrightarrow$ (i).

(i) $\Longrightarrow$ (ii): We have $N_{Y/X} = \mathcal{O}_Y(Y)$ is ample on Y. As in the proof ((ii) and (iii)) of Theorem 5.1, Chapter I, we can conclude (for $n >> 0$) that

a) $|nY|$ has no base points,

b) $H^0(X,\mathcal{O}_X(nY)) \longrightarrow H^0(Y,\mathcal{O}_Y(nY))$ is surjective, and

c) $\mathcal{O}_Y(nY)$ is very ample.

Let $\varphi: X \longrightarrow \mathbb{P}^N_k$ be the morphism defined by the complete linear system $|nY|$. Let $X" = \varphi(X)$ and $Y" = \varphi(Y)$. We know that $\varphi|_Y: Y \longrightarrow Y"$ is an isomorphism, and $Y"$ is a very ample effective divisor on $X"$. Furthermore $Y = \varphi^{-1}(Y")$.

Let $\varphi = g \cdot f$ be the Stein factorization of φ: we have a commutative diagram

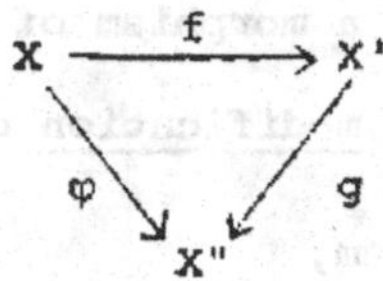

where $f: X \longrightarrow X'$ a proper morphism with $f_*(\mathcal{O}_X) = \mathcal{O}_{X'}$, and $g: X' \longrightarrow X''$ a finite morphism ([EGA,III,4.3.3]). Notice that $f|_Y: Y \xrightarrow{\cong} Y' = f(Y)$ is an isomorphism, and $Y = f^{-1}(Y')$, so by Zariski's Main Theorem ([EGA,III,4.4.1]), f is an isomorphism in a neighborhood of Y onto an open subset of X'. Thus in particular f is birational.

Now we have $Y' = g^{-1}(Y'')$ with g being a finite morphism and Y'' an ample divisor on X''. Hence by Proposition 4.4, Chapter I, we get that Y' is an effective ample divisor on X'. This completes the proof of the theorem.

Remark. Let us analyze what happens to the open set $U = X-Y$ in the above theorem. Assume further X and Y are non-singular. Since rank $N_{Y/X} = 1$, the condition "(i) $N_{Y/X}$ is ample" implies that $N_{Y/X}$ is Γ-ample. Hence by Theorem 4.1 (b), we have $q(U) = 0$. On the other hand the condition (ii) above implies that there exists a proper (and birational) morphism $f: U \longrightarrow U'$ where $U' = X' - Y'$ with $f_*(\mathcal{O}_U) = \mathcal{O}_{U'}$. But U' is affine since Y' is ample on X' (Proposition 2.1, Chapter II). Finally the set $\{y \in U' \mid \dim_{k(y)} f^{-1}(y) > 0\}$ is finite (Exercise 4.12, below). Thus U is a modification of the affine scheme U' in the sense of Goodman-Hartshorne [1]:

Let $f: S \longrightarrow S'$ be a morphism of schemes (of finite type over k). We say that S is a <u>modification</u> of S' (via the map f) if

a) f is a proper morphism,

b) $f_*(\mathcal{O}_S) = \mathcal{O}_{S'}$, and

c) the set $B = \{y \in S' \mid \dim_{k(y)} f^{-1}(y) > 0\}$ is finite.

It follows by [EGA, III, §4], that B is closed in S', and that the restriction of f to $S-f^{-1}(B)$ is an isomorphism onto $S'-B$. In particular f is birational when S and S' are integral.

Naturally enough, by abstracting this situation, we have the

<u>Theorem</u> (<u>Goodman-Hartshorne</u> [1]). Let S be a scheme (of finite type over k). Then $q(S) = 0$ if and only if S is a modification of an affine scheme.

We use this theorem in the following corollary (which is an analogue of Goodman's theorem (Theorem 6.1, Chapter II)).

<u>Corollary 4.3</u>. Let X be a complete variety, and let $Y \subseteq X$ be a closed subset. Let $U = X - Y$. Then U is a modification of an affine scheme if and only if there exists a closed subscheme $Z \subseteq Y$ such that $N_{\overline{Y}/\overline{X}}$ is ample, where $f: \overline{X} \longrightarrow X$ is the birational blowing-up of X with respect to the closed subscheme $Z \subseteq Y$, and $\overline{Y} = f^{-1}(Y)$.

<u>Proof</u>. Suppose U is a modification of an affine scheme, say $f: U \longrightarrow U'$ is a modification map with U' an affine scheme. Since f is birational, f is an isomorphism of $U - W$ on to $U' - f(W)$ where W is closed in U. But W being proper over k, it follows

that W is closed in X. Now define a scheme X' as covered by U' and $X - W$, and glued along $U' - f(W) \cong U - W$. Let $Y' = X' - U'$. Note that $Y \cong Y'$. Write $W' = f(W)$.

Since U' is an affine open subset of X', by Goodman (Theorem 6.1, Chapter II), we have a birational blowing-up (of X' with respect to a closed subscheme $Z' \subseteq Y'$) $g: \overline{X}' \longrightarrow X'$ such that $\overline{Y}' = g^{-1}(Y')$ is an effective ample divisor on $\overline{X}'$.

Let $Z \subseteq Y$ be the subscheme corresponding to $Z' \subseteq Y'$ (since $Y \cong Y'$). Now define a scheme $\overline{X}$ as covered by U and $\overline{X}' - \overline{W}'$, glued along $U - W \cong \overline{U}' - \overline{W}'$ where $\overline{W}' = g^{-1}(W')$. Let $\overline{U}$ denote the open set U in $\overline{X}$. We have $\overline{X} - \overline{U} \cong \overline{Y}'$ and so $N_{\overline{Y}/\overline{X}}$ is ample (where $\overline{Y} = \overline{X} - \overline{U}$). Clearly $\overline{X}$ is the blowing-up of X with respect to Z.

Conversely, if $N_{\overline{Y}/\overline{X}}$ is ample, it follows by Theorem 4.2 (ii), that $q(\overline{U}) = 0$. But $\overline{U} \cong U$ and so $q(U) = 0$. But then by the theorem of Goodman-Hartshorne quoted above, U is a modification of an affine scheme.

<u>Conjecture 4.4</u>. Let X be a complete non-singular variety (over k with $\text{char}\, k = 0$). Let $Y \subseteq X$ be a non-singular subvariety such that $N_{Y/X}$ is ample. Then show that nY (as a cycle) moves in a large algebraic family.

<u>Remark</u>. If $N = N_{Y/X}$ is "sufficiently ample" in the sense that $H^1(Y,N) = 0$ and N is generated by global sections, then Y itself moves in a large algebraic family. This follows from Grothendieck's differential study of the Hilbert scheme [3 , exposé 221, §5].

Conjecture 4.5. Let X be a complete non-singular variety (with char $k = 0$). Suppose $Y,Z \subseteq X$ are non-singular subvarieties such that $N_{Y/X}, N_{Z/X}$ are ample, and $\dim Y + \dim Z \geq \dim X$. Then show that $Y \cap Z \neq \emptyset$. (True if Y or Z is a divisor - Exercise 4.13, below).

Problem 4.6. Let $Y \subseteq \mathbb{P} = \mathbb{P}^n_k$ be a non-singular closed subscheme. Then give a necessary and sufficient conditions so that $N_{Y/\mathbb{P}}$ is Γ-ample.

Exercise 4.7. Let A be a noetherian regular local ring, and let I be an ideal such that A/I is regular. Then show that I is generated by an A-sequence.

Exercise 4.8. Let A be a noetherian local ring, and let I be an ideal. Then show that the following are equivalent.

1) I is generated by an A-sequence,
2) (a) The A/I-module I/I^2 is free, and
   (b) for each n, the canonical surjective homomorphism

$$S^n_{A/I}(I/I^2) \longrightarrow I^n/I^{n+1}$$

is an isomorphism.

Exercise 4.9. Let E be a vector bundle on a scheme X/k with char $k = 0$. Show that $\Gamma^n(E) \cong S^n(E)$. Give an example to show that this need not be true if char $k \neq 0$.

Exercise 4.10 (Gieseker). Let $\mathbb{P} = \mathbb{P}^2_k$ with char $k = p$. Let

$$\mathcal{O}^3_{\mathbb{P}} \xrightarrow{\ f\ } \mathcal{O}_{\mathbb{P}}(2) \longrightarrow 0$$

be an exact sequence of sheaves. Dualize and twist by $\mathcal{O}_{\mathbb{P}}(1)$ to obtain the exact sequence

$$0 \longrightarrow \mathcal{O}_{\mathbb{P}}(-1) \xrightarrow{\check{f}(1)} \mathcal{O}_{\mathbb{P}}(1)^{\oplus 3} \longrightarrow E \longrightarrow 0.$$

Show that E is ample, but not Γ-ample. In fact, if G is any locally free sheaf on $\mathbb{P}$, then

$$H^1(\mathbb{P}, G \otimes \Gamma^{p^n}(E)) \neq 0$$

for all $n \gg 0$.

Exercise 4.11. Let X be a non-singular curve over k of any characteristic. Then a vector bundle E on X is ample if and only if E is Γ-ample.

Exercise 4.12. (Goodman-Hartshorne [1], Lemma 2, p. 261). Let $f\colon S \longrightarrow S'$ be a proper morphism (of schemes of finite type over k). Assume that $H^1(S,F)$ is a finite-dimensional k-vector space for every coherent sheaf F on S, and that S' contains no complete curves. Then show that the set

$$\{y \in S' \mid \dim_{k(y)} f^{-1}(y) > 0\}$$

is finite.

Exercise 4.13. Let X be a complete non-singular variety, and let $Y \subseteq X$ be an effective Cartier divisor with $N_{Y/X}$ ample. Let $Z \subseteq X$ be a closed subscheme of dimension ≥ 1, which is locally a complete intersection with $N_{Z/X}$ ample. Then show that $Y \cap Z \neq \emptyset$.

Exercise 4.14. Subvarieties with negative normal bundles. We say that a vector bundle E on a complete scheme is negative if $E^{\vee}$ is ample. Let X be a complete non-singular variety of dimension n. Let $Y \subseteq X$

be a non-empty closed subscheme which is locally a complete intersection with $N_{Y/X}$ negative. Then show that

$$p(U) = q(U) = cd\ (U) = n-1$$

where $U = X - Y$.

<u>Exercise 4.15</u>. <u>The Albanese Variety</u> Alb X. Let X be a complete non-singular variety. The <u>Albanese variety</u> of X, Alb X, is defined by the following universal property: Fix a point $x_o \in X$. Then Alb X is an abelian variety with a morphism $\varphi: X \longrightarrow \text{Alb } X$ such that

a) $\varphi(x_o) = e$, the identity element of Alb X, and

b) if $\psi: X \longrightarrow B$ is any morphism of X into an abelian variety B such that $\psi(x_o) = e_B$, then there exists a unique homomorphism

$$\rho: \text{Alb } X \longrightarrow B$$

such that $\psi = \rho \cdot \varphi$. (For the existence of Alb X, see for example Chevalley [1], III, pp. 487-489 or Serre [7].

Prove the following result due to Matsumura. If Y is a non-singular subvariety of X of dimension ≥ 1 with $N_{Y/X}$ ample, then the induced homomorphism Alb Y $\longrightarrow$ Alb X is surjective. (Assume for simplicity that the base points $x_o \in X$ and $y_o \in Y$ coincide).

<u>Conjecture 4.16</u>. Let $Y \subseteq X$ be as above. Assume that $\dim Y > \frac{1}{2} \dim X$. Then Alb Y $\longrightarrow$ Alb X is an isomorphism.

§5. Subvarieties of Projective Space.

Let $\mathbb{P} = \mathbb{P}^n_k$ be the projective space of dimension n (k algebraically closed, and of arbitrary characteristic). Let $Y \subseteq \mathbb{P}$ be a closed subscheme of pure dimension s, and let $U = \mathbb{P} - Y$.

The general problem we study in the rest of this chapter is to describe the integers $p(U)$, $q(U)$ and $cd(U)$ in terms of various data on Y. If Y is a set-theoretic complete intersection, we have $p(U) = q(U) = cd(U) = n-s-1$. If Y is non-singular, we have $p(U) = q(U) = n-s-1$, and we have theorems giving an almost complete determination of $cd(U)$ in terms of certain cohomological properties of Y (see statements below). For subvarieties Y with singularities, we have a criterion for $cd(U) < n-1$, but not much else is known, so we state some problems and conjectures. Recall however the theorem of Lichtenbaum (Corollary 3.5) which says that $cd(U) < n \iff Y$ is non-empty.

In our proofs, we deal separately with the cases char $k = 0$ and char $k = p > 0$. The techniques are entirely different, and even the results are not always the same. We will state the results in this section, and refer to the following sections for proofs. We study the case char $k = p > 0$ in §6, the case $k = \mathbb{C}$ in §7, and the case char $k = 0$ in §8.

A closed subscheme $Y \subseteq \mathbb{P}$ is said to be a complete intersection if the largest homogeneous ideal I_Y of Y in $S = k[x_o,\dots,x_n]$ is generated by $\operatorname{codim}(Y,\mathbb{P}) = r$ elements; or, equivalently, if there exist r hypersurfaces H_i such that $Y = \bigcap_{i=1}^{r} H_i$ (as schemes in $\mathbb{P}$).

A closed subset $Y \subseteq \mathbb{P}$ is said to be a <u>set-theoretic complete intersection</u> if Y is the support of a closed subscheme which is a complete intersection.

<u>Theorem 5.1</u>. Suppose Y is a set-theoretic complete intersection of dimension s in $\mathbb{P} = \mathbb{P}^n_k$. Then

$$p(U) = q(U) = \mathrm{cd}\,(U) = n-s-1.$$

<u>Proof</u>. Since Y is pure of dimension s, and is a set-theoretic complete intersection in $\mathbb{P}$, there exist $n-s$ hypersurfaces $H_i \subseteq \mathbb{P}$ such that $Y = \bigcap_{i=1}^{n-s} H_i$. Hence U is the union of the $n-s$ affine open subsets $\mathbb{P} - H_i$. Thus using Čech cohomology and the theorem of Leray ([EGA, III, 1.4.1] or Godement [1], II, 5.4.1), we find that

$$H^i(U,F) = 0$$

for all $i \geq n-s$, and all coherent sheaves F on U. Hence $\mathrm{cd}\,(U) \leq n-s-1$.

On the other hand, by Corollary 3.8, we have $p(U) \geq n-s-1$. Using Lemma 3.7 and Theorem 3.4, we find that

$$p(U) \leq q(U) \leq \mathrm{cd}\,(U)$$

so all three must be equal to $n-s-1$, as required.

Remark. This theorem can be used to show that certain subvarieties of projective space are not set-theoretic complete intersections. For example, there exists a non-singular surface $Y \subseteq \mathbb{P} = \mathbb{P}_k^4$ such that

$$H^2(U,\omega) \neq 0 .$$

See Hartshorne [CDAV], p. 449. For this surface we have $q(U) = 1$ and $cd(U) = 2$, so it cannot be a set-theoretic complete intersection.

Theorem 5.2. Suppose Y is a non-singular subvariety of $\mathbb{P}$. Then

$$p(U) = q(U) = n-s-1.$$

(If char $k \neq 0$, it suffices to assume that Y is Cohen-Macaulay.)

Proof. i) Suppose char $k = 0$. Since Y is non-singular, by Proposition 2.1, $N_{Y/\mathbb{P}}$ is ample, and hence is Γ-ample. But then by Theorem 4.1 (b), we have $p(U) = q(U) = n-s-1$.

ii) Suppose char $k \neq 0$. Assume merely that Y is Cohen-Macaulay. Then the result is Corollary 6.7, below.

Problem 5.3. Let Y be an arbitrary s-dimensional subscheme of $\mathbb{P}$. Show that $p(U) = n-s-1$.

Conjecture 5.4. Fix an integer r (for example $n-s$). Then the following are equivalent.

1) $q(U) \leq r-1$,

2) For every (not necessarily closed) point $y \in Y$.

$$H^i_{J_y}(\mathcal{O}_{\mathbb{P},y}) = 0$$

for all $i > r$ where J_y is the ideal of Y in $\mathcal{O}_{\mathbb{P},y}$,

3) Let $S = k[x_o,\ldots,x_n]$ be the homogeneous coordinate ring of $\mathbb{P}$. Then the graded S-module

$$M_j = \sum_{m\in\mathbb{Z}} H^j(\hat{\mathbb{P}},\hat{F}(m))$$

is an S-module of finite type for all $j < n-r$, and all locally free sheaves F on $\mathbb{P}$.

Remarks. i) The implication $(3)\Longrightarrow(1)$ follows from Theorem 3.4(a). Otherwise nothing is known.

ii) If Y is a local complete intersection, then (2) is satisfied for $r = n-s$. (Exercise 5.9, below). Thus this conjecture would imply a conjecture of Grothendieck [SGA 2], exposé XIII, Conjecture 1.3.

iii) If Y is Cohen-Macaulay, and char $k = p > 0$, then (1) is satisfied for $r = n-s$ (Theorem 5.2).

iv) If Y is a set-theoretic complete intersection, then (1), (2) and (3) are all satisfied for $r = n-s$ (see Exercise 5.10, below).

Here we state the main results on cd (U) which will be proved in §§6, 7, and 9, below.

(i) The case characteristic p.

Theorem (See Corollaries 6.8 and 6.10, below). Let Y be a Cohen-Macaulay subscheme of $\mathbb{P} = \mathbb{P}_k^n$ with char $k = p > 0$. Let r be an integer. Assume that codim $(Y,\mathbb{P}) \leq r \leq n-1$. Then the following are equivalent.

1) cd $(U) < r$,

2) Y is connected, and $H^i(Y,\mathcal{O}_Y)_s = 0$ for $0 < i < n-r$ (where the

subscript s denotes the stable part of $H^i(Y,\mathcal{O}_Y)$ in its Fitting decomposition),

3) $H^i(U,\omega) = 0$ for $i \geq r$, where $\omega = \Omega^n_{U/k}$.

(ii) The case $k = \mathbb{C}$, the field of complex numbers.

Theorem (See Theorem 7.4, and Corollary 2.4 of Chapter VI, below). Let Y be a closed subscheme of $\mathbb{P} = \mathbb{P}^n_{\mathbb{C}}$. Let r be an integer. Then $\operatorname{cd}(U) < r$ implies that the natural maps

$$H^i(\mathbb{P}^h, \mathbb{C}) \longrightarrow H^i(Y^h, \mathbb{C})$$

are isomorphisms for $i < n-r$, and injective for $i = n-r$. Conversely, if Y is non-singular of dimension s, and $r \geq n-s$, then the latter condition implies $\operatorname{cd}(U) < r+1$. Furthermore (for non-singular Y) we always have $\operatorname{cd}(U) < 2n-2s$.

(iii) The case characteristic zero.

For an arbitrary field of characteristic 0, we can give a purely algebraic proof of one half of the previous theorem, using algebraic De Rham cohomology instead of ordinary complex cohomology.

Theorem (See Theorem 8.6, below). Let Y be a closed subscheme of $\mathbb{P} = \mathbb{P}^n_k$ with $\operatorname{char} k = 0$. Let r be an integer. Then $\operatorname{cd}(U) < r$ implies that the natural maps

$$H^i_{DR}(\mathbb{P}) \longrightarrow H^i_{DR}(Y)$$

are isomorphisms for $i < n-r$, and injective for $i = n-r$.

(iv) Subvarieties with singularities (in any characteristic).

For subvarieties Y with arbitrary singularities we have the following result (in any characteristic).

Theorem (See Theorem 3.2, Chapter V). Let Y be a closed subscheme of $\mathbb{P} = \mathbb{P}^n_k$. Then the following are equivalent.

1) Y is connected and of dimension ≥ 1,

2) $\mathrm{cd}(U) < n-1$.

(See also Corollary 6.9, below, for a different proof in the case char $k \neq 0$).

In connection with this theorem, we pose the classical

Problem 5.5. Suppose Y is a connected (even irreducible) curve in $\mathbb{P}^3_k$. Then is it true that Y is a set-theoretic complete intersection?

Remark. A necessary condition for the curve Y to be a set-theoretic complete intersection is $\mathrm{cd}(U) = 1$ (Theorem 5.1). But this condition is satisfied for any connected curve, by the theorem just stated. Hence the problem must be attacked by some other means.

By the usual analogy between statements on projective space and statement about local rings, we are led to the following conjecture about local rings. First we make a definition.

Let A be a noetherian local ring with residue field k. We say that an A-module N is cofinite if $\mathrm{Supp}(N) = V(\mathfrak{m})$, where $\mathfrak{m}$ = maximal ideal of A, and $\mathrm{Hom}_A(k,N)$ is a finite-dimensional k-vector space. See Grothendieck [LC], §§4 and 6, for some equivalent conditions. See also Hartshorne [6], §1.

Conjecture 5.6. Let A be a regular local ring of dimension n with residue field $k = A/\mathfrak{m}$. Let J be an ideal in A, and let r be a fixed integer. Then the following are equivalent.

1) $H^i_J(A)$ is cofinite for all $i > r$,

2) For every prime ideal $\mathfrak{p} \neq \mathfrak{m}$,

$$H^i_{JA_{\mathfrak{p}}}(A_{\mathfrak{p}}) = 0$$

for all $i > r$.

Remark. Note that the second condition is satisfied for $r = \operatorname{codim} V(J)$ if $V(J)$ has an isolated singularity at $\mathfrak{m}$. Even in tnat case, it is not known if the first condition is true.

Conjecture 5.7. (Local analogue of Theorem 3.2, Chapter V). Let A be a complete regular local ring of dimension n, and let $J \subseteq \mathfrak{m}$ be an ideal such that $V(J)$ has dimension ≥ 2, and $V(J) - \{\mathfrak{m}\}$ is connected. Then

$$H^{n-1}_J(A) = 0.$$

Exercise 5.8. Let Y be a non-empty closed subset of $\mathbb{P} = \mathbb{P}^n_k$, and let $U = \mathbb{P} - Y$. Then show that $p(U) \leq q(U)$.

Exercise 5.9. Let A be a noetherian local ring, and let J be an ideal generated by r elements. Assume that $\dim (A/J) = \dim A - r$. Then show that $H^i_J(A) = 0$ for $i > r$.

Exercise 5.10. Let Y be a set-theoretic complete intersection in $\mathbb{P} = \mathbb{P}_k^n$ with $\dim Y = s$. Show that the graded $S = k[x_o,\ldots,x_n]$-module

$$M^j = \sum_{m\in\mathbb{Z}} H^j(\hat{\mathbb{P}}, \hat{F}(m))$$

is of finite type for all $j < s$, and all locally free sheaves F on $\mathbb{P}$.

Exercise 5.11. Let Y be a curve in $\mathbb{P} = \mathbb{P}_k^3$, and assume that there is a surface $Z \subseteq \mathbb{P}$ with $Y \subseteq Z$ and $Z - Y$ affine. Then use the local cohomology sequence for $Z - Y$ in $\mathbb{P} - Y$ to show that $H^2(\mathbb{P} - Y, F) = 0$ for all coherent sheaves F on $\mathbb{P}$. Thus $\mathrm{cd}\,(\mathbb{P} - Y) = 1$. This gives a simpler proof of Theorem 3.2, Chapter V, in this special case.

Exercise 5.12. Let $\mathbb{P} = \mathbb{P}_k^4$ with homogeneous coordinates $x_o,\ldots,x_4$. Let $Y = Y_1 \cup Y_2$ where $Y_1 = \{x_1 = x_2 = 0\}$ and $Y_2 = \{x_3 = x_4 = 0\}$. Thus Y_1 and Y_2 are projective planes which meet in the single point $P = (1,0,0,0,0)$. Using a Mayer-Vietoris sequence for local cohomology, show that for any non-zero locally free sheaf F on $\mathbb{P}$

$$H^i(U,F) = \begin{cases} 0 \quad \text{for} \quad i \geq 3 \\ \infty\text{-dimensional for} \quad i = 1,2 \\ \text{finite-dimensional for} \quad i = 0 \end{cases}.$$

More precisely, since $H^2(U,F)$ is the group of interest for calculating $q(U)$ and $\mathrm{cd}\,(U)$, show that $H^2(U, \mathcal{O}_{\mathbb{P}}) \cong I_P$, where I_P is the injective hull of k over the local ring $\mathcal{O}_{\mathbb{P},P}$. Thus it is the bad point P which gives rise to the infinite-dimensional cohomology group. We have $p(U) = 1$; $q(U) = \mathrm{cd}\,(U) = 2$.

It follows that the subset Y of this example is <u>not</u> a set-theoretic complete intersection.

<u>Exercise 5.13</u>. Let X be a non-singular projective variety. Let $Y \subseteq X$ be a closed subscheme of dimension s. Suppose Y is a set-theoretic complete intersection. Assume that $Y = Y_1 \cup Y_2$ where Y_1 and Y_2 are unions of irreducible components of Y, and no irreducible component of Y is contained in $Y_1 \cap Y_2$. Then prove that

$$\dim\,(Y_1 \cap Y_2) = s-1.$$

<u>Exercise 5.14</u>. Let $Y \subseteq \mathbb{P} = \mathbb{P}^n_k$ be a closed subset having no isolated points. Prove that $H^{n-1}(U,F)$ is finite-dimensional for all coherent sheaves F.

<u>Exercise 5.15</u>. Let Y be the rational cubic curve in $\mathbb{P}^3_k$ defined as the image of the morphism

$$\varphi\colon \mathbb{P}^1_k \longrightarrow \mathbb{P}^3_k$$

given by $(t,u) \longmapsto (t^3,t^2u,tu^2,u^3)$. Show that the reduced curve Y is not a complete intersection (in the strict sense) but it is a set-theoretic complete intersection.

<u>Exercise 5.16</u>. Let Y be the non-singular rational quartic curve in $\mathbb{P}^3_k$ defined as the image of the morphism

$$\varphi\colon \mathbb{P}^1_k \longrightarrow \mathbb{P}^3_k$$

given by $(t,u) \longmapsto (t^4,t^3u,tu^3,u^4)$. Show that if char $k = p \neq 0$,

Y is a set-theoretic complete intersection (for any p). We do not know if Y is a set-theoretic complete intersection in characteristic zero.

Consideration of the cone over a curve in $\mathbb{P}_k^3$ leads us to the following conjecture in local algebra.

Conjecture 5.17. Let A be a regular local ring containing a field of characteristic zero. Let B be a local ring which is a finitely generated flat A-module. Then B_{red} is also a flat A-module.

Exercise 5.18. Show that this conjecture, if true, would imply that the rational quartic curve of Exercise 5.16, is not a set-theoretic complete intersection in characteristic zero. At the same time, Exercise 5.16 shows that the conjecture is false in char $k = p \neq 0$. There are other counter-examples to this conjecture in char. p due to M. Nagata and C.P. Ramanujam.

Remark. Closely related to the question of set-theoretic complete intersections is the weaker question: if Y is a subset of $\mathbb{P}_k^n$, what is the least number of hypersurfaces needed to cut out Y set-theoretically? One recent result is the theorem of M. Kneser [1], which says that any irreducible curve in $\mathbb{P}_k^3$ is a set-theoretic intersection of at most three hypersurfaces.

Problem 5.19. Generalize Kneser's result to higher dimensions. Is it true, for example, that any connected subvariety $Y \subseteq \mathbb{P}_k^n$ of dimension s is a set-theoretic intersection of at most $2n-2s-1$ hypersurfaces? (This would imply a theorem of Barth, see §1 (v), Chapter VI).

§6. Subvarieties of Projective Space in characteristic $p > 0$.

In this section we work over an algebraically closed field k of characteristic $p > 0$. Let Y be a closed subscheme of the projective space $\mathbb{P} = \mathbb{P}^n_k$, and let $U = \mathbb{P} - Y$. Our problem is to calculate $cd(U)$ in terms of some data on Y. Using the results of §3 above, we reduce to studying the cohomology of the formal completion $\hat{\mathbb{P}}$ of $\mathbb{P}$ along Y. The special feature of characteristic p is that we have the Frobenius morphism, which allows us to calculate the cohomology groups of $\hat{\mathbb{P}}$ in terms of the cohomology groups of Y, and the action of Frobenius. This is our main result (Theorem 6.2, below).

We begin with some preliminaries of a purely algebraic nature. Then after the theorem we derive some corollaries which give information about $cd(U)$.

i) The Fitting decomposition.

Let V be a finite-dimensional vector space over k. An additive endomorphism f of V is said to be a *p-linear endomorphism* *if*

$$f(\alpha v) = \alpha^p f(v)$$

for all $v \in V$ and all $\alpha \in k$.

A p-linear endomorphism f of V induces a *unique* decomposition, called the *Fitting decomposition* of V as

$$V = V_s \oplus V_n$$

where V_s and V_n are subspaces of V such that $f|_{V_s}$ (resp. $f|_{V_n}$)

is an automorphism (resp. nilpotent). The subspace V_s (resp. V_n) is called the <u>stable</u> (resp. <u>nilpotent</u>) part of V with respect to f.

Since k is perfect, $f^m(V)$ is a subspace of V for every m. Since V is finite-dimensional, we have

$$f^m(V) = f^{m+1}(V) = \ldots = V_s \text{ (say)}$$

and

$$\ker f^m = \ker f^{m+1} = \ldots = V_n \quad \text{(say)} ,$$

for $m \gg 0$. Clearly $f|_{V_s}$ (resp. $f|_{V_n}$) is an automorphism (resp. nilpotent), and

$$V_s \cap V_n = (0) .$$

Now given a $v \in V$, there exists a $w \in V$ such that $f^m(v) = f^{2m}(w)$. We have $f^m(w) \in V_s$ and $v - f^m(w) \in V_n$. But $v = f^m(w) + v - f^m(w)$. Thus $V = V_s \oplus V_n$. Uniqueness is clear.

ii) <u>The Frobenius Endomorphism</u> π.

Let Y be a scheme of finite type over k. We define a new scheme Y_p over k by taking the same scheme $(Y, \mathcal{O}_Y)$, but the k-algebra structure of $\mathcal{O}_Y$ being given by multiplying by the pth roots of elements of k. We define a k-morphism (called the <u>Frobenius morphism</u>)

$$\pi: Y \longrightarrow Y_p$$

by giving the identity map on the spaces, and the pth power map on the structure sheaves. Note that since Y is of finite type over k, π is a finite (obviously surjective) morphism.

For any coherent sheaf F on Y, denote by F_p the same sheaf considered on Y_p. We can describe $\pi^*(F_p)$ as a tensor product with an explanation: $\pi^*(F_p) = F \otimes_{\mathcal{O}_Y} \mathcal{O}_Y$, where F is an $\mathcal{O}_Y$-module in the usual way; $\mathcal{O}_Y$ is an $\mathcal{O}_Y$-module via the pth power map, and the tensor product is an $\mathcal{O}_Y$-module by multiplying in the factor on the right in the usual way.

Since no confusion is likely, we will drop the suffix p in Y_p (resp. F_p) and write simply $\pi\colon Y \longrightarrow Y$ (resp. $\pi^*(F)$). For example, we have the induced p-linear homomorphisms

$$\pi^*\colon\ H^i(Y,F) \longrightarrow H^i(Y, \pi^*(F))$$

for all i.

Now let Y be a closed subscheme of $\mathbb{P} = \mathbb{P}^n_k$. Then if $F = \mathcal{O}_Y(\nu)$ with $\nu \in \mathbb{Z}$, we have $\pi^*(F) = \mathcal{O}_Y(p\nu)$, and hence we have p-linear maps

$$\pi^*\colon H^i(Y, \mathcal{O}_Y(\nu)) \longrightarrow H^i(Y, \mathcal{O}_Y(p\nu))$$

for all i. For each i, denote by

$$M^i = \sum_{\nu\in\mathbb{Z}} H^i(Y, \mathcal{O}_Y(\nu)).$$

We know that each M^i is canonically a graded S-module where $S = k[X_o,\ldots,X_n]$ is the homogeneous coordinate ring of the projective space $\mathbb{P}$. Now the p-linear map π^* induces an endomorphism of the graded group M^i for each i. In fact, this map defines the structure of a graded (S,F)-module on each M^i in the following sense.

iii) Graded (S,F)-modules.

Let $S = k[X_o,\ldots,X_n]$. We will denote by $F: S \longrightarrow S$ the pth power endomorphism of the ring S: $F(s) = s^p$.

By a graded (S,F)-module (or simply an (S,F)-module), we mean a pair (M,f) where M is a graded S-module, and $f: M \longrightarrow M$ is an endomorphism of graded groups satisfying

a) $f(sm) = s^p f(m)$

for all $s \in S$ and all $m \in M$, and

b) $\deg(f(m)) = p.\deg(m)$

for all homogeneous elements $m \in M$. We will denote by M_ν the k-vector space of homogeneous elements of degree ν of M, for $\nu \in \mathbb{Z}$.

Clearly (S,F) is an (S,F)-module. For an (S,F)-module (M,f), it is clear that $f|_{M_o}$ is a p-linear endomorphism of the k-vector space M_o. So we can speak of its Fitting decomposition, provided M_o is finite-dimensional (for example when M is a finite type S-module).

By a homomorphism of (S,F)-modules (M,f) and (M',f'), we mean a homogeneous homomorphism (of degree zero) $g: M \longrightarrow M'$ of graded S-modules such that the diagram

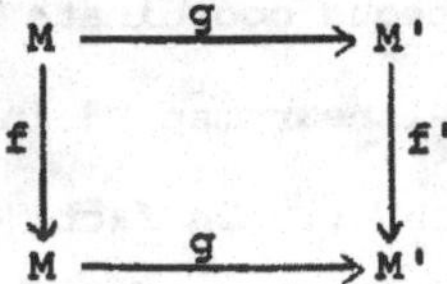

is commutative. It is clear that the graded (S,F)-modules form a category which we denote by (S,F)-mod.

Examples. 1. If Y is a closed subscheme of $\mathbb{P} = \mathbb{P}_k^n$, then the graded S-module

$$M^i = \sum_{\nu \in \mathbb{Z}} H^i(Y, \mathcal{O}_Y(\nu)),$$

together with the Frobenius endomorphism $\pi^*: M^i \longrightarrow M^i$, is a graded (S,F)-module.

2. With the same notations, let ^ denote the formal completion along Y. Then for each i, the graded S-module

$$\sum_{\nu \in \mathbb{Z}} H^i(\hat{\mathbb{P}}, \mathcal{O}_{\hat{\mathbb{P}}}(\nu)),$$

together with the Frobenius endomorphism π^*, is an (S,F)-module.

3. If (M,f) is an (S,F)-module, and if V is a k-vector space with a p-linear endomorphism g, then $(M \otimes_k V, f \otimes g)$ is an (S,F)-module.

4. We denote by (S,p^r) the graded S-module S, with the usual action on the left, and with action by p^r-th powers on the right. If $s \in S$, we denote the corresponding element of the module (S,p^r) by (s,p^r). Thus (S,p^r) is a bimodule satisfying

$$t \cdot (s,p^r) = (ts,p^r)$$

$$(s,p^r) \cdot t = (st^{p^r},p^r)$$

for s,t S. If we let f be the pth power endomorphism of S, then $((S,p^r),f)$ is an (S,F)-module, on both sides.

iv) <u>The Functor</u> G.

We will now define a functor

$$G: (S,F)\text{-}\underline{\text{mod}} \longrightarrow (S,F)\text{-}\underline{\text{mod}}$$

as follows. Let (M,f) be an (S,F)-module. Consider the tensor product

$$(S,p^r) \otimes_S M$$

of the right S-module (S,p^r) and the left S-module M. This is a graded group under the grading

$$\deg((s,p^r)\otimes m) = \deg(s) + p^r \cdot \deg(m)$$

for all homogeneous elements $s \in S$ and $m \in M$. This graded group is considered as an S-module via the left S-module structure on (S,p^r). The endomorphism

$$F\otimes f: (S,p^r) \otimes M \longrightarrow (S,p^r) \otimes M$$

clearly makes $(S,p^r)\otimes M$ into an (S,F)-module.

We define a homomorphism of (S,F)-modules

$$\varphi_{r+1,r}: (S,p^{r+1})\otimes M \longrightarrow (S,p^r)\otimes M$$

by $\varphi_{r+1,r} = id\otimes f$, i.e., $(s,p^{r+1})\otimes m \longmapsto (s,p^r)\otimes f(m)$. Then the maps $\varphi_{r+1,r}$ make the set of modules $\{(S,p^r)\otimes M\}_{r\geq 0}$ into an inverse system of (S,F)-modules.

For an inverse system $\{N^\alpha,\varphi_{\alpha,\beta}\}$ of graded modules (over some graded ring) where the $\varphi_{\alpha,\beta}$ are homogeneous homomorphisms of degree zero, we define the <u>restricted inverse limit</u>, denoted $\varprojlim_\alpha{}' (N^\alpha)$, by

$$\varprojlim_{\alpha}{}' (N^{\alpha}) = \sum_{\nu\in\mathbb{Z}} \varprojlim_{\alpha} (N^{\alpha}_{\nu}) .$$

This is again canonically a graded module.

It is a matter of simple verification that the restricted inverse limit

$$\varprojlim_{r}{}' ((S,p^{r}) \otimes M)$$

is in fact a graded (S,F)-module. Now define G by

$$G(M) = \varprojlim_{r}{}' ((S,p^{r}) \otimes M).$$

It is easily checked that G is then a functor on the category of (S,F)-modules.

The following theorem helps us to calculate $G(M)$ in certain cases.

<u>Theorem 6.1</u>. Let M be an (S,F)-module. Assume that $M_{\nu} = 0$ for $\nu \ll 0$ and M_{0} is finite-dimensional over k. Then

$$G(M) \approx S \otimes_{k} (M_{0})_{s}$$

as (S,F)-modules, where $(M_{0})_{s}$ denotes the stable part of M_{0} in its Fitting decomposition. (This result applies in particular when M is of finite type over S).

<u>Proof</u>. For any $\mu \in \mathbb{Z}$, we have (by definition)

$$G(M)_{\mu} = \varprojlim_{r} ((S,p^{r}) \otimes M)_{\mu} .$$

By definition of the gradation on $(S,p^{r}) \otimes M$, we have

$$((S,p^r)\otimes M)_\mu = (S,p^r)_\mu \otimes_k M_o + \sum_{\nu\neq 0} (S,p^r)_{\mu-p^r\nu} \otimes_k M_\nu$$

$$= (S,p^r)_\mu \otimes_k M_o + \sum_{\nu<0} (S,p^r)_{\mu-p^r\nu} \otimes_k M_\nu$$

for $r >> 0$ (for example if $p^r > \mu$). By definition of the homomorphism

$$\varphi_{r+t,r}: ((S,p^{r+t}) \otimes M)_\mu \longrightarrow ((S,p^r) \otimes M)_\mu ,$$

we find that

$$\varphi_{r+t,r}((S,p^{r+t})_\mu \otimes_k M_o) \subseteq (S,p^r)_\mu \otimes_k M_o$$

and

$$\varphi_{r+t,r}(\sum_{\nu<0} (S,p^{r+t})_* \otimes_k M_\nu) \subseteq \sum_{\nu<0} (S,p^r)_* \otimes_k M_{p^t\nu}$$

$$= 0$$

for $t >> 0$, since $M_\nu = 0$ for $\nu << 0$. Hence it follows that

$$G(M)_\mu = \varprojlim_r ((S,p^r)_\mu \otimes_k M_o)$$

$$= S_\mu \otimes_k \varprojlim_r (M_o)$$

$$= S_\mu \otimes_k (M_o)_s .$$

Thus

$$G(M) = \sum_\mu S_\mu \otimes_k (M_o)_s$$

$$= S \otimes_k (M_o)_s .$$

It is easy to see that this is an isomorphism of (S,F)-modules. This completes the proof.

Now we return to the geometric situation. Most of our results on cohomological dimension in characteristic p are consequences of the following theorem.

Theorem 6.2. Let Y be a closed subscheme of the projective space $\mathbb{P} = \mathbb{P}^n_k$ over an algebraically closed field k of characteristic $p > 0$. Let $\wedge$ denote the formal completion along Y. Then for each $i \geq 0$ there is a canonical isomorphism of graded (S,F)-modules.

$$G(\sum_{m\in\mathbb{Z}} H^i(Y,\mathcal{O}_Y(m))) \xrightarrow{\ \cong\ } \sum_{m\in\mathbb{Z}} H^i(\hat{\mathbb{P}},\mathcal{O}_{\hat{\mathbb{P}}}(m))$$

where $S = k[x_0,\ldots,x_n]$, and G is the functor defined above.

Proof. Let I be the sheaf of ideals defining Y. For each r, let Y_r be the subscheme defined by the ideal $I^{(p^r)}$ generated by the p^r-th powers of elements of I. These ideals define the same topology on $\mathcal{O}_{\mathbb{P}}$ as the ordinary powers of I. Hence we can write

$$H^i(\hat{\mathbb{P}},\mathcal{O}_{\hat{\mathbb{P}}}(m)) \cong \varprojlim H^i(Y_r,\mathcal{O}_{Y_r}(m)),$$

and

$$\sum_{m\in\mathbb{Z}} H^i(\hat{\mathbb{P}},\mathcal{O}_{\hat{\mathbb{P}}}(m)) \cong \varprojlim{}' \sum_{m\in\mathbb{Z}} H^i(Y_r,\mathcal{O}_{Y_r}(m))$$

where $\varprojlim{}'$ is the restricted inverse limit as defined earlier.

Let $\pi\colon \mathbb{P} \longrightarrow \mathbb{P}$ be the Frobenius morphism, and let π^r be its r-fold iteration. Then π^r induces the p^r-th power homomorphism on the structure sheaf $\mathcal{O}_{\mathbb{P}}$. Therefore $(\pi^r)^*\mathcal{O}_Y = \mathcal{O}_{Y_r}$. Thus we can write

$$\mathcal{O}_{Y_r}(m) = (\pi^r)^*\mathcal{O}_Y \otimes \mathcal{O}_{\mathbb{P}}(m).$$

Applying the finite morphism π^r, and using the projection formula (Exercise 6.12, below) we have isomorphisms

$$H^i(Y_r, \mathcal{O}_{Y_r}(m)) \cong H^i(Y, \mathcal{O}_Y \otimes \pi^r_* \mathcal{O}_{\mathbb{P}}(m)) .$$

Thus to establish the isomorphism of the theorem, it will be sufficient to establish isomorphisms

$$\varphi_r : (S, p^r) \otimes_S \sum_{\mu} H^i(Y, \mathcal{O}_Y(\mu)) \xrightarrow{\sim} \sum_{\lambda} H^i(Y, \mathcal{O}_Y \otimes \pi^r_* \mathcal{O}_{\mathbb{P}}(\lambda))$$

of graded (S,F)-modules, for each r such that the isomorphisms φ_r are compatible with the maps in the inverse systems on each side.

We define the map φ_r as follows. Let $s \in S$ be an element of degree ν, and let $m \in H^i(Y, \mathcal{O}_Y(\mu))$. Then we can consider s as an element of $H^0(\mathbb{P}, \pi^r_* \mathcal{O}_{\mathbb{P}}(\nu))$. Note by the projection formula (Exercise 6.12, below) that

$$\mathcal{O}_Y(\mu) \otimes \pi^r_* \mathcal{O}_{\mathbb{P}}(\nu) \cong \pi^r_* \mathcal{O}_{\mathbb{P}}(\nu + p^r\mu).$$

We define φ_r by the cup-product:

$$\varphi_r(s \otimes m) = s \cup m \in H^i(Y, \mathcal{O}_Y \otimes \pi^r_* \mathcal{O}_{\mathbb{P}}(\nu + p^r\mu)) .$$

It is a matter of trivial (but tedious) verification to see that φ_r is a morphism of graded (S,F)-modules, and that for varying r, the morphisms are compatible with the maps in the two inverse systems.

To complete the proof, we need some lemmas.

<u>Lemma 6.3</u>. Let E be a coherent sheaf on $\mathbb{P} = \mathbb{P}^n_k$. Then there exist integers m_i such that $E \cong \sum_i \mathcal{O}_{\mathbb{P}}(m_i)$ if and only if

i) $H^0(\mathbb{P}, E(m)) = 0$ for all $m \ll 0$, and

ii) $H^i(\mathbb{P}, E(m)) = 0$ for all m and $0 < i < n$.

<u>Proof</u>. Suppose $E = \sum_i \mathcal{O}_{\mathbb{P}}(m_i)$ for some m_i. Then (i) and (ii) are consequences of [EGA, III, 2.1.12] and the fact that $H^0(\mathbb{P}, \mathcal{O}_{\mathbb{P}}(m)) = 0$ for $m < 0$.

Conversely, suppose (i) and (ii) are true. Then by (i), the graded $S = k[x_0, \ldots, x_n]$-module

$$M = \sum_{m \in \mathbb{Z}} H^0(\mathbb{P}, E(m))$$

is of finite type over S. Indeed, by [EGA, III, 2.3.2], it satisfies the condition TF, which says that for some m_0, the submodule $M' = \sum_{m \geq m_0} M_m$ is of finite type. On the other hand, $M_m = 0$ for $m \ll 0$ by (i), and each M_m is finite-dimensional over k. Hence M is of finite type.

Let $\mathfrak{M} = (x_0, \ldots, x_n)$ be the maximal ideal in S. Then there is a natural exact sequence

$$0 \longrightarrow H^0_{\mathfrak{M}}(M) \longrightarrow M \longrightarrow \sum_m H^0(\mathbb{P}, E(m)) \longrightarrow H^1_{\mathfrak{M}}(M) \longrightarrow 0$$

and there are isomorphisms for each $i > 0$

$$\sum_m H^i(\mathbb{P}, E(m) \cong H^{i+1}_{\mathfrak{M}}(M).$$

Indeed, the same statement occurs in Grothendieck [EGA, III, 2.1.5], with $H^i((\underline{x}),M)$ in place of $H^i_{\mathfrak{M}}(M)$. But these groups are isomorphic (see [EGA, III, 1.1.10] or Grothendieck [LC], 2.3).

Now from the definition of M, and from our hypotheses, we find that $H^i_{\mathfrak{M}}(M) = 0$ for all $i < n+1$. So by Grothendieck [LC], Corollary 3.10, we have $\text{depth}_{\mathfrak{M}} M = n+1$. (It could not be bigger than n+1, because $\dim S = n+1$). Since M is an S-module of finite type and $S_{\mathfrak{M}}$ is a regular local ring of dimension n+1, by [EGA, 0_{IV}, 17.3.4], we have

$$\text{depth}_{\mathfrak{M}}(M) + \text{hd}_{\mathfrak{M}}(M) = n+1.$$

where $\text{hd}_{\mathfrak{M}}(M)$ denotes the homological dimension of the $S_{\mathfrak{M}}$-module $M_{\mathfrak{M}}$. Thus we get

$$\text{hd}_{\mathfrak{M}}(M) = 0$$

which means that $M_{\mathfrak{M}}$ is free. Now since M is a graded S-module, by Exercise 6.16, below, M is a free S-module and hence

$$M = \sum_i S(m_i)$$

for suitable integers m_i (= - degree of the generators of M). This gives that

$$E = \sum_i \mathcal{O}_{\mathbb{P}}(m_i).$$

Corollary 6.4. Suppose $\pi\colon \mathbb{P} \longrightarrow \mathbb{P}$ is the Frobenius morphism. Then for each $m \in \mathbb{Z}$, there are integers $m_i \in \mathbb{Z}$ such that

$$\pi_*(\mathcal{O}_{\mathbb{P}}(m)) \approx \sum_i \mathcal{O}_{\mathbb{P}}(m_i)$$

Proof. It suffices to verify the condition (i) and (ii) of the above lemma for $E = \pi_*(\mathcal{O}_{\mathbb{P}}(m))$. We have

$$H^i(\mathbb{P},\pi_*(\mathcal{O}_{\mathbb{P}}(m)) = H^i(\mathbb{P},\mathcal{O}_{\mathbb{P}}(m))$$

$$= 0 \quad \text{for} \quad 0 < i < n \quad \text{and all} \quad m.$$

Also by the projection formula, we have

$$H^0(\mathbb{P},\pi_*(\mathcal{O}_{\mathbb{P}}(m))(m')) = H^0(\mathbb{P},\pi_*(\mathcal{O}_{\mathbb{P}}(m+pm'))$$

$$= 0 \quad \text{for} \quad m' \ll 0.$$

<u>Lemma 6.5</u>. Let $\mathbb{P}$ be a projective space. Let E be a sheaf of the form $\sum \mathcal{O}_{\mathbb{P}}(m_i)$, and let F be any coherent sheaf on $\mathbb{P}$. Then there are natural isomorphisms of graded S-modules.

$$\sum_{\mu} H^0(\mathbb{P}, E(\mu)) \otimes_S \sum_{\nu} H^i(\mathbb{P}, F(\nu)) \xrightarrow{\sim} \sum_{\lambda} H^i(\mathbb{P}, E \otimes F(\lambda)).$$

<u>Proof</u>. The map is given by the cup-product. To prove it is an isomorphism, we may assume $E \approx \mathcal{O}_{\mathbb{P}}(m)$. Then the result is trivial, because

$$\sum_{\mu} H^0(\mathbb{P}, E(\mu)) \cong S(m),$$

the free S-module on one generator in degree $-m$.

<u>Proof of theorem</u>, <u>continued</u>. Take $F = \mathcal{O}_Y$, $E = \pi^r_* \mathcal{O}_{\mathbb{P}}(m)$ in the last lemma. The sheaf E is of the form $\sum \mathcal{O}_{\mathbb{P}}(m_i)$ by the previous corollary, so the lemma applies. We get an isomorphism

$$\sum_{\mu} H^0(\mathbb{P},(\pi^r_*\mathcal{O}_{\mathbb{P}}(m))(\mu)) \otimes_S \sum_{\nu} H^i(Y,\mathcal{O}_Y(\nu)) \xrightarrow{\sim} \sum_{\lambda} H^i(Y,\mathcal{O}_Y \otimes(\pi^r_*\mathcal{O}_{\mathbb{P}}(m)(\lambda)) .$$

Now $(\pi^r_*\mathcal{O}_{\mathbb{P}}(m))(\mu) = \pi^r_*(\mathcal{O}_{\mathbb{P}}(m+p^r\mu))$. So this isomorphism can be written

$$\sum_{\mu} (S,p^r)_{m+p^r\mu} \otimes_S \sum_{\nu} H^i(Y,\mathcal{O}_Y(\nu)) \xrightarrow{\sim} \sum_{\lambda} H^i(Y,\mathcal{O}_Y \otimes \pi^r_* \mathcal{O}_{\mathbb{P}}(m+p^r\lambda)).$$

Now letting m vary in the range $0 \leq m < p^r$ and summing, we get the required isomorphism φ_r.

For the following corollaries, we fix our notation. Let $\mathbb{P} = \mathbb{P}^n_k$ be projective space over an algebraically closed field k of characteristic $p > 0$. Let Y be a closed subscheme of $\mathbb{P}$, and let $U = \mathbb{P} - Y$.

Corollary 6.6. Assume for some i that $H^i(Y,\mathcal{O}_Y(m)) = 0$ for all $m \ll 0$. Then there is an isomorphism of graded (S,F)-modules

$$\sum_{m \in \mathbb{Z}} H^i(\hat{\mathbb{P}}, \mathcal{O}_{\hat{\mathbb{P}}}(m)) \cong S \otimes_k H^i(Y,\mathcal{O}_Y)_s ,$$

where the subscript s denotes the stable part of $H^i(Y,\mathcal{O}_Y)$ in its Fitting decomposition.

Proof. This follows immediately from Theorems 6.1 and 6.2.

Corollary 6.7. Assume that Y is Cohen-Macaulay of dimension s. Then $p(U) = q(U) = n-s-1$.

Proof. Since Y is Cohen-Macaulay, we have $H^i(Y,\mathcal{O}_Y(m)) = 0$ for $m \ll 0$ and $i < s$ (Exercise 6.13, below). Thus the previous corollary applies, and we find that $H^i(\hat{\mathbb{P}},\mathcal{O}_{\hat{\mathbb{P}}}(m))$ is finite-dimensional for all m, and for all $i < s$. (In fact, it is 0 for $m < 0$.) On the other hand, $H^i(\hat{\mathbb{P}}, \mathcal{O}_{\hat{\mathbb{P}}}(m)) = 0$ for $i > s$, since $\mathcal{O}_{\mathbb{P}}(m)$ has support on Y, which is of combinatorial dimension s (Grothendieck [1, Theorem 3.6.5]. Thus by Lemma 3.7, we must have $H^s(\hat{\mathbb{P}},\mathcal{O}_{\hat{\mathbb{P}}}(-s))$

infinite-dimensional. Now it follows from Theorem 3.4 that $p(U) = q(U) = n-s-1$.

Corollary 6.8. Assume that Y is Cohen-Macaulay, and let r be an integer satisfying $\operatorname{codim}(Y,\mathbb{P}) \leq r \leq n-1$. Then the following conditions are equivalent:

(i) $\operatorname{cd}(U) < r$,

(ii) Y is connected, and $H^i(Y,\mathcal{O}_Y)_s = 0$ for $0 < i < n-r$.

Proof. We have already seen that $\operatorname{cd}(U) < n-1$ implies that Y is connected (Corollary 3.9). Note by Proposition 3.1 that we can find the cohomological dimension of U using only locally free sheaves. In fact it is sufficient to use the sheaves $\mathcal{O}_{\mathbb{P}}(m)$, $m \in \mathbb{Z}$. Thus $\operatorname{cd}(U) < r$ is equivalent, by Theorem 3.4, to saying that the natural maps

$$\alpha_i\colon\ H^i(\mathbb{P}, \mathcal{O}_{\mathbb{P}}(m)) \longrightarrow H^i(\hat{\mathbb{P}}, \mathcal{O}_{\hat{\mathbb{P}}}(m))$$

are isomorphisms for $i < n-r$, and injective for $i = n-r$, for all m $\mathbb{Z}$. For $i > 0$, we have $H^i(\mathbb{P}, \mathcal{O}_{\mathbb{P}}(m)) = 0$ for all m. Thus we must see if $H^i(\hat{\mathbb{P}},\mathcal{O}_{\hat{\mathbb{P}}}(m)) = 0$ for $0 < i < n-r$. Since Y is Cohen-Macaulay, we can apply Corollary 6.6. Thus $H^i(\hat{\mathbb{P}}, \mathcal{O}_{\hat{\mathbb{P}}}(m)) = 0$ for $0 < i < n-r$ and all m if and only if $H^i(Y,\mathcal{O}_Y)_s = 0$ for $0 < i < n-r$. To complete the proof, we need only observe that if Y is connected then α_0 is an isomorphism. Indeed, we may assume that Y is reduced. Corollary 6.6 applies for $i = 0$ for any Y without isolated points. If Y is connected, we have $H^0(Y,\mathcal{O}_Y) \cong k$, and so

$$\sum_m H^0(\hat{\mathbb{P}}, \mathcal{O}_{\hat{\mathbb{P}}}(m)) \cong S ,$$

and α_o is an isomorphism.

As a special case of this result, we have

<u>Corollary 6.9</u>. Let Y be any subscheme of $\mathbb{P}$. Then cd $(U) < n-1$ if and only if Y is connected and of dimension ≥ 1.

<u>Corollary 6.10</u>. Assume that Y is Cohen-Macaulay, and let r be an integer satisfying codim $(Y,\mathbb{P}) \leq r \leq n-1$. Then the following conditions are equivalent:

(i) cd $(U) < r$,

(ii) $H^i(U,\omega) = 0$ for $i \geq r$, where $\omega = \Omega^n_{U/k}$.

<u>Proof</u>. By Corollary 6.6, the maps

$$\alpha_i: H^i(\mathbb{P}, \mathcal{O}_{\mathbb{P}}(m)) \longrightarrow H^i(\hat{\mathbb{P}}, \mathcal{O}_{\hat{\mathbb{P}}}(m))$$

are isomorphisms for all m and $i < n-r$ if and only if $H^o(\hat{\mathbb{P}}, \mathcal{O}_{\hat{\mathbb{P}}}) \cong k$ and $H^i(\hat{\mathbb{P}}, \mathcal{O}_{\hat{\mathbb{P}}}) = 0$ for $0 < i < n-r$. Now by the technique of proof of Theorem 3.4, this is equivalent to $H^i(U,\omega) = 0$ for $i \geq r$.

<u>Remark</u>. In other words, to find the cohomological dimension of U, we need only test the single sheaf ω.

Exercises

Exercise 6.11. Let $S = k[x_o,\ldots,x_n]$. Let I be an injective hull of k over S (see Grothendieck [LC], §4). We can describe I explicitly as follows: I is the k-vector space generated by the expressions

$$\left\{ x_o^{i_o}\ldots x_n^{i_n} \middle| i_j < 0, \quad j = 0,\ldots,n \right\} .$$

The action of S is as one would expect, except that every time any exponent becomes ≥ 0, one gets 0. Prove that $G(I) \cong I$.

Exercise 6.12. Prove the projection formula: Let $f\colon X \longrightarrow Y$ be an affine morphism of noetherian schemes. Let F be a quasi-coherent sheaf on X, and let G be a quasi-coherent sheaf on Y. Then there is a natural functorial isomorphism $f_*(F)\otimes_Y G \xrightarrow{\sim} f_*(F \otimes_X f^*G)$. (See [EGA, 0_I, 5.4.10]).

Exercise 6.13. Let Y be a projective Cohen-Macaulay scheme, and let F be a locally free sheaf on Y. Show that

$$H^i(Y,F(m)) = 0 \quad \text{for} \quad m \ll 0 \quad \text{and} \quad i < \dim Y.$$

(This is a generalization of the lemma of Enriques-Severi-Zariski. See Zariski [3] and Serre [FAC], §76).

Exercise 6.14. Let $\mathcal{A}$ be the subcategory of the category of (S,F)-modules consisting of those (S,F)-modules M such that M_ν is finite-dimensional for all $\nu \in \mathbb{Z}$. (Here M_ν denotes the homogeneous elements of degree ν in M.) Show that G is an exact functor from $\mathcal{A} \longrightarrow (S,F)$-mod.

Exercise 6.15. Show that the implication (i) $\Longrightarrow$ (ii) of Corollary 6.8 holds even without the hypothesis Y Cohen-Macaulay.

Exercise 6.16. Let $S = k[x_o,\ldots,x_n]$, and let $\mathfrak{M} = (x_o,\ldots,x_n)$. Let M be a graded S-module of finite type. If $M_{\mathfrak{M}}$ is a free $S_{\mathfrak{M}}$-module then prove that M is a free S-module.

§7. Transcendental results in characteristic 0.

In this section we will consider varieties over the complex numbers $\mathbb{C}$. If X is a complete variety, containing a subvariety Y, we will show that $cd(X-Y)$ being small implies a certain connection between the (usual) complex cohomology of X and Y (Theorem 7.4, below). Our technique is to use a theorem of Grothendieck on the algebraic De Rham cohomology, and then some standard topological results.

First we will define the algebraic De Rham cohomology. Let S be a non-singular variety of dimension n over a field k (of arbitrary characteristic). Let $\Omega^1_{S/k}$ be the sheaf of differential 1-forms of S over k. It is a locally free sheaf of rank n, since S is non-singular. We denote by Ω^i the sheaf $\Lambda^i\Omega^1_{S/k}$ of differential i-forms, for $i = 1,2,\ldots,n$. The natural derivation $d\colon \mathcal{O}_S \longrightarrow \Omega^1_S$ gives rise to k-linear maps $d\colon \Omega^i \longrightarrow \Omega^{i+1}$ for each i. The complex

$$\mathcal{O}_S \xrightarrow{d} \Omega^1 \xrightarrow{d} \Omega^2 \xrightarrow{d} \cdots \xrightarrow{d} \Omega^n$$

is called the <u>algebraic De Rham complex</u> of S, and is denoted by $\Omega^{\cdot}_S$. The hypercohomology of this complex ([EGA, 0_{III}, 11.4.3]) gives the <u>algebraic De Rham cohomology</u>

$$H^i_{DR}(S) = H^i(S,\Omega^{\cdot}_S).$$

<u>Proposition 7.1</u>. There is a spectral sequence

$$E_1^{pq} = H^q(S,\Omega^p) \Longrightarrow E^n = H^n_{DR}(S) .$$

Proof. This is just the spectral sequence of hypercohomology ([EGA, 0_{III}, 11.4.3.1]).

Remark. This proposition shows that the De Rham cohomology groups are finite-dimensional if S is proper over k. For in that case the terms E_1^{pq} of the spectral sequence are all finite-dimensional. On the other hand, it follows from the theorem of Grothendieck below that the groups $H_{DR}^i(S)$ are again finite-dimensional if $k = \mathbb{C}$. In characteristic p, however, the groups $H_{DR}^i(S)$ need not be finite-dimensional (see Exercise 7.7, below). For another proof of finite-dimensionality in characteristic 0, see Monsky [2].

Proposition 7.2. Let S be a non-singular scheme of dimension n over k and assume that cd (S) < r, for some integer r. Then $H_{DR}^i(S) = 0$ for $i \geq n + r$.

Proof. In the spectral sequence of Proposition 7.1, we have $E_1^{pq} = 0$ for $p > n$ (since dim S = n) and for $q \geq r$ (since cd S < r). Therefore in the abutment, we have $H_{DR}^i(S) = 0$ for $i \geq n+r$.

The usefulness of the algebraic De Rham cohomology derives mostly from the following theorem of Grothendieck, which compares the algebraic De Rham cohomology of S with the usual complex cohomology of the associated complex analytic space S^h. (See Chapter VI, §2, below, for the notion of associated analytic space.)

Theorem (Grothendieck [7], Theorem 1'). Let S be a non-singular scheme of finite type over $\mathbb{C}$. Then there are natural isomorphisms

$$H^i_{DR}(S) \xrightarrow{\cong} H^i(S^h,\mathbb{C})$$

for all i, where the groups on the right are the singular cohomology groups of the associated analytic space S^h, with complex coefficients.

We will also need the following theorem in topology.

Theorem 7.3. Let X be a compact polyhedron, and let Y be a closed subpolyhedron, and assume that $X-Y$ is an n-manifold. Then

$$H^i(X,Y;\mathbb{C}) \cong H^{n-i}(X-Y,\mathbb{C})' .$$

Proof. We refer to Spanier [1] for the topological results we need. First note that

$$H^i(X,Y;\mathbb{C}) \cong \bar{H}^i(X,Y;\mathbb{C})$$

since X and Y are compact polyhedra (Spanier, p. 291). For the definition of $\bar{H}^i$ see Spanier, p. 289. Now by Lefschetz duality (Spanier, p. 297) we have

$$\bar{H}^i(X,Y;\mathbb{C}) \cong H_{n-i}(X-Y;\mathbb{C}) .$$

Finally, by the universal coefficient theorem (Spanier, p. 244), we have

$$H_{n-i}(X-Y;\mathbb{C}) \cong H^{n-i}(X-Y,\mathbb{C})' ,$$

since all the groups in question are finitely generated.

Now we can state our main result.

Theorem 7.4. Let X be a complete scheme of dimension n over $\mathbb{C}$. Let Y be a closed subscheme, and assume that X-Y is non-singular. Let r be an integer. Then cd (X-Y) < r implies that the natural maps

$$H^i(X^h,\mathbb{C}) \longrightarrow H^i(Y^h,\mathbb{C})$$

are isomorphisms for $i < n-r$, and injective for $i = n-r$.

Proof. Since X-Y is non-singular, we can consider its algebraic De Rham cohomology. By Proposition 7.2, cd (X-Y) < r implies that $H^i_{DR}(X-Y) = 0$ for $i \geq n+r$. Now by the theorem of Grothendieck above, $H^i(X^h-Y^h,\mathbb{C}) = 0$ for $i \geq n+r$. Hence by Theorem 7.3,

$$H^i(X^h,Y^h;\mathbb{C}) = 0 \quad \text{for} \quad i \leq n-r ,$$

since X^h-Y^h is a 2n-manifold. Now finally, our result follows from the exact sequence of relative cohomology (Spanier [1], p. 240)

$$\cdots \longrightarrow H^i(X^h,Y^h;\mathbb{C}) \longrightarrow H^i(X^h,\mathbb{C}) \longrightarrow H^i(Y^h,\mathbb{C}) \longrightarrow H^{i+1}(X^h,Y^h;\mathbb{C}) \rightarrow \cdots$$

Corollary 7.5. Let Y be a non-singular subvariety of $\mathbb{P} = \mathbb{P}^n_{\mathbb{C}}$. Then cd ($\mathbb{P}$-Y) < r implies

$$H^q(Y,\Omega^p) = \begin{cases} 0 & \text{for } p \neq q,\ p+q < n-r \\ \mathbb{C} & \text{for } p = q,\ p+q < n-r . \end{cases}$$

Proof. Since Y is projective and non-singular, we have

$$H^i(Y^h,\mathbb{C}) \cong \sum_{p+q=i} H^q(Y^h,(\Omega^p)^h)$$

by the theorem of Dolbeault [1]. On the other hand, by a theorem of Serre [GAGA], Theorem 1, p. 19, we have isomorphisms

$$H^q(Y,\Omega^p) \xrightarrow{\ \cong\ } H^q(Y^h,(\Omega^p)^h) .$$

Now the maps of cohomology

$$H^i(\mathbb{P}^h,\mathbb{C}) \longrightarrow H^i(Y^h,\mathbb{C})$$

are compatible with the Dolbault decomposition, and the cohomology groups $H^q(\mathbb{P},\Omega^p)$ are known to be 0 for $p \neq q$, and $\mathbb{C}$ for $p = q$ (see Exercise 7.8, below). Thus the result follows from the theorem.

Remark. The fact that $\operatorname{cd}(\mathbb{P} - Y) < n-2 \Longrightarrow H^1(Y,\Theta_Y) = 0$ was proved by another method by Hartshorne [CDAV], Theorem 8.5, p. 448. Note that $\operatorname{Pic}^o Y = 0 \Longrightarrow H^1(Y,\Theta_Y) = 0$.

Corollary 7.6. Let Y be a non-singular subvariety of dimension s of $\mathbb{P} = \mathbb{P}^n_{\mathbb{C}}$, and assume that Y is a set-theoretic complete intersection. Then

$$H^q(Y,\Omega^p) = \begin{cases} 0 & \text{for } p \neq q,\ p+q < s \\ \mathbb{C} & \text{for } p = q,\ p+q < s . \end{cases}$$

Proof. In that case, $\operatorname{cd}(\mathbb{P} - Y) = n-s-1$ (Proposition 5.8, Chapter III).

Remark. This generalizes a theorem of Kodaira and Spencer [1]. See also Chapter IV, §4, for a general account of Lefschetz-type theorems.

Exercise 7.7. Let $X = \operatorname{Spec}(k[x])$ where k is a field of characteristic $p > 0$. Show that $H^i_{DR}(X)$ is infinite-dimensional for $i = 0,1$.

<u>Exercise 7.8</u>. Let $\mathbb{P} = \mathbb{P}^n_k$, the characteristic of k being arbitrary. Show that

$$H^q(\mathbb{P}, \Omega^p_{\mathbb{P}/k}) = \begin{cases} 0 & \text{for } p \neq q \\ k & \text{for } p = q,\ 0 \leq p \leq n. \end{cases}$$

Deduce that

$$H^i_{DR}(\mathbb{P}) = \begin{cases} 0 & \text{for } i \text{ odd} \\ k & \text{for } i \text{ even},\ 0 \leq i \leq 2n. \end{cases}$$

§8. The Gysin sequence for algebraic De Rham cohomology.

In this section we give a purely algebraic proof of Theorem 7.4, using algebraic De Rham cohomology instead of complex cohomology. The proof is valid over any algebraically closed field of characteristic zero. Our main tool is a Gysin sequence for algebraic De Rham cohomology. We also prove a Poincaré duality theorem for this cohomology.

Theorem 8.1. Let X be a non-singular variety over an algebraically closed field k of characteristic 0. Let Y be a non-singular closed subvariety of codimension r. Then there is a natural morphism of complexes

$$\varphi\colon \Omega_Y^{\cdot} \longrightarrow \underline{H}_Y^r(\Omega_X^{\cdot})[r] ,$$

which induces an isomorphism of the cohomology sheaves of these complexes. (We say φ is a quasi-isomorphism, in the terminology of Hartshorne [RD], Chapter I, §4). Here $\Omega_Y^{\cdot}$ (resp. $\Omega_X^{\cdot}$) is the algebraic De Rham complex on Y (resp. X), which was defined in section 7. The symbol $[r]$ means "shift r places left".

Proof. First we must define the map φ. Since X and Y are both non-singular, we have an exact sequence of sheaves on Y

$$0 \longrightarrow I/I^2 \longrightarrow \Omega_X^1 \otimes \mathcal{O}_Y \longrightarrow \Omega_Y^1 \longrightarrow 0 \quad ([\text{EGA,IV,17.2.5}]).$$

Now I/I^2 is locally free of rank r, so for each i, we have a natural map

$$\Lambda^r(I/I^2) \otimes \Lambda^i\Omega_Y^1 \longrightarrow \Lambda^{i+r}(\Omega_X^1 \otimes \mathcal{O}_Y) .$$

If we define

$$\omega_{Y/X} = \Lambda^r(I/I^2)^{\vee} ,$$

This gives us a natural map,

$$\Omega_Y^i \longrightarrow \Omega_X^{i+r} \otimes \omega_{Y/X} . \qquad (1)$$

Now by the "<u>fundamental local isomorphism</u>" (Hartshorne [RD], Chapter III, 7.2), there is a canonical isomorphism

$$\Omega_X^{i+r} \otimes \omega_{Y/X} \cong \underline{\mathrm{Ext}}^r_{\mathcal{O}_X}(\mathcal{O}_Y, \Omega_X^{i+r}) . \qquad (2)$$

But

$$\underline{H}^r_Y(\Omega_X^{i+r}) \cong \varinjlim_n \underline{\mathrm{Ext}}^r_{\mathcal{O}_X}(\mathcal{O}_X/I_Y^n, \Omega_X^{i+r})$$

(see Grothendieck [LC], 2.8), so there is a natural map

$$\underline{\mathrm{Ext}}^r_{\mathcal{O}_X}(\mathcal{O}_Y, \Omega_X^{i+r}) \longrightarrow \underline{H}^r_Y(\Omega_X^{i+r}) . \qquad (3)$$

Combining (1), (2), and (3), we obtain for each i a map

$$\varphi^i : \Omega_Y^i \longrightarrow \underline{H}^r_X(\Omega_X^{i+r}) .$$

Clearly these commute with the differentiation maps d in each complex, so the φ^i give us a morphism of complexes, φ.

To show that φ is a quasi-isomorphism is a purely local question. Hence we may assume that $X = \mathrm{Spec}\, A$, where A is a regular local ring, and that Y is given by equations $x_1 = \ldots = x_r = 0$, where $x_1,\ldots,x_r$ is part of a regular system of parameters in A.

We will use induction on r to reduce to the case $r = 1$. Let Z be defined by $x_1 = 0$. Then we have $Y \subseteq Z \subseteq X$. Suppose inductively the maps

$$\varphi_{Y/Z}: \Omega_Y^{\cdot} \longrightarrow \underline{H}_Y^{r-1}(\Omega_Z^{\cdot})[r-1]$$

and

$$\varphi_{Z/X}: \Omega_Z^{\cdot} \longrightarrow \underline{H}_Z^{1}(\Omega_X^{\cdot})[1]$$

are quasi-isomorphisms. Since Y is complete intersection in Z and Z is complete intersection in X, we have

$$\underline{H}_Y^{r-1}(\underline{H}_Z^1(\Omega_X^{\cdot})) \cong \underline{H}_Y^r(\Omega_X^{\cdot})$$

(see Exercise 8.7, below). Therefore by composition, we find that

$$\varphi_{Y/X} = \underline{H}_Y^{r-1}(\varphi_{Z/X}) \cdot \varphi_{Y/Z}$$

is a quasi-isomorphism.

Now we may assume $r = 1$. In this case we write $Y = \operatorname{Spec} B$, where $B = A/xA$, and x is a regular parameter in A. For any coherent sheaf F on X, we have the long exact sequence of local cohomology

$$H^0(X,F) \longrightarrow H^0(X-Y,\ F) \longrightarrow H_Y^1(X,F) \longrightarrow 0.$$

Applying this to the sheaves Ω_A^i, and noting that $X-Y = \operatorname{Spec} A\ [\frac{1}{x}]$, we obtain an isomorphism

$$H_Y^1(\Omega_A^i) \cong \Omega_A^i[\tfrac{1}{x}]/\Omega_A^i\ .$$

Here we use Ω_A^i to denote the A-module as well as the sheaf. Since X is affine, we may deal with modules over A instead of sheaves on X. Thus φ becomes a morphism of complexes of A-modules whose

i^{th} component is

$$\varphi^i : \Omega^i_B \longrightarrow \Omega^{i+1}_A[\tfrac{1}{x}]/\Omega^{i+1}_A .$$

With these identifications, φ^i can be described explicitly as follows. For any $\eta \in \Omega^i_B$,

$$\varphi^i(\eta) = \eta \wedge \frac{dx}{x} .$$

Let $\hat{A}$ be the completion of A with respect to the x-adic topology. The modules Ω^i_B and $\Omega^{i+1}_A[\frac{1}{x}]/\Omega^{i+1}_A$ both have a natural structure of $\hat{A}$-module, so we may replace A by $\hat{A}$.

Now we need the following lemma of Grothendieck.

Lemma. Let A be a noetherian ring, containing a field k, and complete with respect to an x-adic topology, where x is a non-zero divisor in A. Let $B = A/xA$, and assume that B is smooth over k. Then there is an isomorphism $A \cong B[[x]]$.

Proof. Apply Grothendieck [SGA 1], exposé III, Corollary 5.6, with $S = \mathrm{Spec}\ k[[x]]$, $X = \mathrm{Spec}\ B[[x]]$, and $Y = \mathrm{Spec}\ A$.

Now to prove that φ is a quasi-isomorphism, we borrow a calculation of Atiyah and Hodge [1], p. 77-78. Having fixed the isomorphism $\hat{A} \cong B[[x]]$, any element

$$\gamma \in \Omega^{i+1}_A[\tfrac{1}{x}]/\Omega^{i+1}_A$$

can be written uniquely in the form

$$\gamma = \frac{\alpha_1}{x} + \dots + \frac{\alpha_s}{x^s} + (\frac{\beta_1}{x} + \dots + \frac{\beta_s}{x^s})dx$$

for suitable $s \geq 0$, with $\alpha_j \in \Omega_B^{i+1}$ and $\beta_j \in \Omega_B^i$ for each j. If $d\gamma = 0$, we find that

$$\gamma = d\Theta + \alpha_1 \frac{dx}{x} \quad \text{and} \quad d\alpha_1 = 0 ,$$

where

$$\Theta = -\frac{\alpha_2}{x} - \frac{\alpha_3}{2x^2} - \cdots - \frac{\alpha_s}{(s-1)x^{s-1}} .$$

(Here we use char $k = 0$.) Secondly, note that

$$\alpha_1 \frac{dx}{x} = d\psi$$

for some ψ if and only if $\alpha_1 = d\rho$ for some ρ. Indeed, ψ must be of the form $\rho \frac{dx}{x}$ for some ρ, and so $\alpha_1 = d\rho$.

Thus φ induces an isomorphism of the closed/exact forms of $\Omega_B^{\cdot}$ onto the closed/exact forms of $\Omega_A^{\cdot}[\frac{1}{x}]/\Omega_A^{\cdot}$, which was to be proved.

Corollary 8.2. Under the hypotheses of the theorem, there is a natural isomorphism, for each i,

$$H^i(Y,\Omega_Y^{\cdot}) \xrightarrow{\cong} H_Y^{i+2r}(X,\Omega_X^{\cdot}) .$$

Proof. A quasi-isomorphism of complexes induces an isomorphism on hypercohomology groups. Hence by the theorem, we have an isomorphism, for each i,

$$H^i(Y,\Omega_Y^{\cdot}) \xrightarrow{\cong} H^i(X, \underline{H}_Y^r(\Omega_X^{\cdot})[r]).$$

Now for any locally free sheaf F on X, the sheaves $\underline{H}_Y^i(F)$ are 0 for $i \neq r$, since Y is locally complete intersection in X (see Exercise 8.7, below). Hence the spectral sequence of composite functors $\Gamma_Y = \Gamma \cdot \underline{\Gamma}_Y$ degenerates (see Grothendieck [LC], 1.4) and we have isomorphisms

$$H^i(X, \underline{H}_Y^r(F)) \cong H_Y^{i+r}(X,F) .$$

Applying this to the sheaves Ω_X^i, and taking into account the shift in degrees, we have

$$H^i(X, \underline{H}_Y^r(\Omega_X^{\cdot})[r]) \cong H_Y^{i+2r}(X,\Omega_X^{\cdot}) .$$

Combining with the above gives the Corollary.

<u>Theorem 8.3</u> (<u>Gysin sequence</u>). Let X be a non-singular variety over an algebraically closed field k of characteristic 0. Let Y be a non-singular closed subvariety of codimension r. Then there is a long exact sequence of algebraic De Rham cohomology

$$\cdots \longrightarrow H_{DR}^i(X) \longrightarrow H_{DR}^i(X-Y) \xrightarrow{\delta} H_{DR}^{i+1-2r}(Y) \longrightarrow H_{DR}^{i+1}(X) \longrightarrow \cdots$$

<u>Proof</u>. We have only to take the long exact sequence of local cohomology (see Grothendieck [LC], §1) of the complex $\Omega_X^{\cdot}$, and use the isomorphisms of the above Corollary.

<u>Remark</u>. The coboundary map δ in the above exact sequence merits the name <u>generalized Poincaré residue</u>. It is best known in the case when X is projective and Y is a hypersurface (see Leray [1]). In that case X-Y is affine, so δ sends a closed regular differential i-form on X-Y to an element of the cohomology class $H_{DR}^{i-1}(Y)$. If the form η has only a simple pole along Y, then $\delta\eta$ can be represented by a closed (i-1)-form ρ on Y, via the edge-homomorphism

$$\ker\, (H^o(Y,\Omega_Y^{i-1}) \xrightarrow{d} H^o(Y,\Omega_Y^i)) \longrightarrow H_{DR}^{i-1}(Y)$$

of the spectral sequence of Proposition 7.1. In particular, if X is a curve and Y a point, we recover the usual notion of residue.

Next we will establish a duality theorem for algebraic De Rham cohomology. We will need a lemma.

Lemma 8.4. Let X be a non-singular projective variety of dimension n over an algebraically closed field of arbitrary characteristic. Then the natural map

$$d: H^n(X,\Omega^{n-1}) \longrightarrow H^n(X,\Omega^n)$$

is the zero map.

Proof. We give separate proofs for char $k = 0$ and char $k = p > 0$. If char $k = 0$, let X be embedded in a projective space $\mathbb{P} = \mathbb{P}^N_k$. Then the end of the Gysin sequence for X in $\mathbb{P}$ gives an exact sequence

$$H^{2n}_{DR}(X) \longrightarrow H^{2N}_{DR}(\mathbb{P}) \longrightarrow H^{2N}_{DR}(\mathbb{P} - X).$$

But $cd(\mathbb{P}-X) < N$ by Lichtenbaum's theorem (Corollary 3.5). Hence $H^{2N}_{DR}(\mathbb{P}-X) = 0$ by Proposition 7.2. On the other hand, $H^{2N}_{DR}(\mathbb{P}) \cong k$ (see Exercise 7.8, above). So we conclude that $H^{2n}_{DR}(X) \neq 0$. Since $H^n(X,\Omega^n) \cong k$ by Serre duality, the map d of the lemma must be zero. Otherwise, in the spectral sequence of Proposition 7.1, we would get $H^{2n}_{DR}(X) = 0$.

If char $k = p > 0$, let $\pi: X \longrightarrow X$ be the Frobenius morphism. Then for each i, we have p-linear isomorphisms

$$H^i(X,\Omega^{\cdot}_X) \cong H^i(X, \pi_*\Omega^{\cdot}_X) .$$

The maps $\pi_* d$ in the complex $\pi_* \Omega_X^{\cdot}$ are $\mathcal{O}_X$-linear. The Cartier operator (see Seshadri [1]) gives isomorphisms

$$(\ker \pi_* d^i)/(\mathrm{im}\, \pi_* d^{i-1}) \cong \Omega_X^i .$$

Thus there is a second spectral sequence of De Rham cohomology [EGA, 0_{III}, 11.4.3.2]

$$E_2^{pq} = H^p(X,\Omega^q) \Longrightarrow H_{DR}^*(X) .$$

In this spectral sequence, we have

$$H^n(X,\Omega^n) = E_2^{nn} \cong H_{DR}^{2n}(X) .$$

Thus $H_{DR}^{2n}(X) \cong k$, and as before, we conclude that the d of the lemma is zero.

Remarks. For $k = \mathbb{C}$, one knows from Hodge theory that all the maps

$$d_1^{pq}\colon\ H^q(X,\Omega^p) \longrightarrow H^q(X,\Omega^{p+1})$$

are zero. In characteristic $p > 0$, these maps need not all be zero (see Mumford [2]).

Theorem 8.5. (Poincaré duality). Let X be a non-singular projective variety over an algebraically closed field k of arbitrary characteristic. Then the cup-product induces perfect pairings, for each i,

$$H_{DR}^i(X) \times H_{DR}^{2n-i}(X) \longrightarrow H_{DR}^{2n}(X) \cong k .$$

<u>Proof</u>. We will work with the spectral sequence

$$E_1^{pq} = H^q(X,\Omega^p) \Longrightarrow H_{DR}^*(X)$$

of Proposition 7.1. The cup-product gives us pairings at each level

$$E_r^{pq} \times E_r^{st} \longrightarrow E_r^{p+s,q+t}$$

and on the abutment

$$H_{DR}^i(X) \times H_{DR}^j(X) \longrightarrow H_{DR}^{i+j}(X).$$

One verifies easily the formula

$$d(a \cup b) = da \cup b - a \cup db$$

for any $a \in E_r^{pq}$, $b \in E_r^{st}$, where $d = d_r$.

Now from the lemma, we have $d_1^{n-1,n} = 0$, and all other maps into E_r^{nn}, for any r, are 0. It follows that the maps d_r commute with the cup-product pairings into E_r^{nn}: for each p,q,r we have a commutative diagram

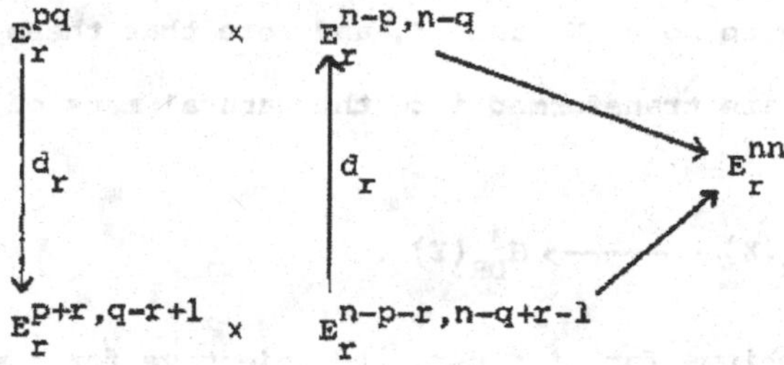

By Serre duality, we have $H^n(X,\Omega^n) \cong k$, and the pairings

$$E_1^{pq} \times E_1^{n-p,n-q} \longrightarrow E_1^{nn} \cong k$$

are perfect. It follows that $E_r^{nn} \cong k$ for all r, and that the pairings into E_r^{nn} are also perfect, for each r. Passing to the limit, we find $H^{2n}_{DR}(X) \cong k$, and the pairings of the theorem are also perfect.

Now we come to our main result.

Theorem 8.6. Let X be a non-singular projective variety over an algebraically closed field k of characteristic 0. Let Y be a non-singular closed subvariety of X, and let r be an integer. Then $\operatorname{cd}(X-Y) < r$ implies that the natural maps

$$H^i_{DR}(X) \longrightarrow H^i_{DR}(Y)$$

are isomorphisms for $i < n-r$, and injective for $i = n-r$.

Proof. By Proposition 7.2, $\operatorname{cd}(X-Y) < r$ implies that $H^i_{DR}(X-Y) = 0$ for $i \geq n+r$. Therefore by the Gysin sequence, the maps

$$H^{i-2r}_{DR}(Y) \longrightarrow H^i_{DR}(X)$$

are isomorphisms for $i > n+r$, and surjective for $i = n+r$. We now apply Poincaré duality to both X and Y, and note that these maps of the Gysin sequence are transformed into the natural maps of cohomology

$$H^j_{DR}(X) \longrightarrow H^j_{DR}(Y) .$$

Thus these are isomorphisms for $j < n-r$, and injective for $j = n-r$.

Exercise 8.7. Let X be a Cohen-Macaulay scheme, and let Y be a closed subscheme of codimension r, which is a local complete intersection. For any locally free sheaf F on X, show that

a) $\underline{H}^i_Y(F) = 0$ for $i \neq r$,

b) $\underline{H}^r_Y(F)$ has a filtration whose quotients are locally free sheaves on Y, and

c) if Z is a closed subscheme of Y of codimension s, which is locally a complete intersection in Y, then

$$\underline{H}^i_Z(\underline{H}^r_Y(F)) = 0 \quad \text{for} \quad i \neq s$$

and

$$\underline{H}^s_Z(\underline{H}^r_Y(F)) \cong \underline{H}^{r+s}_Z(F) .$$

CHAPTER IV

THE GROTHENDIECK-LEFSCHETZ THEOREMS

In this chapter we will prove two theorems of Grothendieck [SGA 2], exposé XII, inspired by Lefschetz (see §4 below for a historical note). The first (Corollary 2.2) states that the fundamental group of a projective algebraic variety X is isomorphic to the fundamental group of a complete intersection subvariety Y of X, provided that $\dim Y \geq 2$. The second (Corollary 3.2) states that the Picard group of a complete intersection Y in $\mathbb{P}^n_k$ is a free abelian group generated by the class of $\mathcal{O}_Y(1)$, provided that $\dim Y \geq 3$.

Grothendieck's method of proof makes essential use of nilpotent elements. From Y he passes through the infinitesimal neighborhoods of Y to the formal completion $\hat{X}$ of X along Y. Then from $\hat{X}$ he passes to a neighborhood U of Y in X, and thence to X itself. The most difficult step is the passage from $\hat{X}$ to U, and for this purpose he introduces the "Lefschetz conditions" on a subvariety (§1 below).

We have restricted our attention to the case of a non-singular ambient variety X. This permits a substantial simplification of Grothendieck's proof. The key result (Theorem 1.5) is that a complete intersection subvariety Y of dimension ≥ 2 satisfies the effective Lefschetz condition. Although suggested by Grothendieck's work, this result was not proved by him.

On the other hand, Grothendieck's more difficult methods also give some local results on purity and factoriality of local rings, which we will not deal with here (see [SGA 2], exposés X and XI).

§1. The Lefschetz Conditions.

We recall the definitions given by Grothendieck ([SGA 2], exposé X, p. 112). Let X be a scheme, and let $Y \subseteq X$ be a closed subscheme . Let ^ denote the formal completion along Y. We say that the pair (X,Y) satisfies the Lefschetz condition, written Lef (X,Y), if for every open set $U \supseteq Y$, and every locally free sheaf E on U, there exists an open set U' with $Y \subseteq U' \subseteq U$ such that the natural map

$$H^0(U',E|_{U'}) \xrightarrow{\cong} H^0(\hat{X},\hat{E})$$

is an isomorphism.

We say that the pair (X,Y) satisfies the effective Lefschetz condition, written Leff (X,Y), if Lef (X,Y), and in addition, for every locally free sheaf $\mathcal{E}$ on $\hat{X}$, there exists an open set $U \supseteq Y$ and a locally free sheaf E on U such that $\hat{E} \cong \mathcal{E}$.

Note that the conditions Lef (X,Y) and Leff (X,Y) are both local around Y.

Remarks. Let GLF (Y) denote the category of germs of locally free sheaves around Y. In other words, an object E of GLF (Y) is a class of locally free sheaves defined on open neighborhoods of Y under the usual equivalence: if E and F are locally free sheaves defined on open neighborhoods U and V of Y respectively, then $E \sim F$ is there exists an open set W with $Y \subseteq W \subseteq U \cap V$ such that $E|_W \cong F|_W$. Morphisms are defined in a natural way.

Let LF $(\hat{X})$ denote the category of locally free sheaves on $\hat{X}$.

Now we have a functor

$$\wedge : \underline{\text{GLF}}\ (Y) \longrightarrow \underline{\text{LF}}\ (\hat{X})$$

sending $E \longmapsto \hat{E}$.

1. The condition Lef (X,Y) implies that this functor is <u>fully faithful</u>.

Indeed, let E and F be two locally free $\mathcal{O}_U$-modules (for some open neighborhood U of Y). Then $\underline{\text{Hom}}\ (E,F)$ is also locally free. By Lef (X,Y), we have an isomorphism of

$$H^o(U', \underline{\text{Hom}}\ (E,F)) \xrightarrow{\ \cong\ } H^o(\hat{X}, (\underline{\text{Hom}}\ (E,F))^\wedge\ )$$

for some open set U' with $Y \subseteq U' \subseteq U$. But $(\underline{\text{Hom}}\ (E,F))^\wedge = \underline{\text{Hom}}\ (\hat{E},\hat{F})$ since E and F are locally free. This proves the assertion.

2. The condition Leff (X,Y) implies that the functor $\wedge$ is an <u>equivalence of categories</u> (obvious).

<u>Proposition 1.1</u>. Let X be a non-singular projective variety of dimension n, and let $Y \subseteq X$ be a closed subscheme. Then the following are equivalent.

(i) cd $(X-Y) < n-1$,

(ii) Lef (X,Y) and Y meets every effective Cartier divisor on X.

<u>Proof</u>. (i) $\Longrightarrow$ (ii). Let $U \supseteq Y$ be any open set, and let E be a locally free sheaf on U. Let $\omega = \Omega^n_{X/k}$. Consider the locally free sheaf $F = \underline{\text{Hom}}_{\mathcal{O}_U}(E,\omega|_U)$. Extend F to a coherent sheaf (denoted again by) F on X. Denote by abuse of notation, $E = \underline{\text{Hom}}_{\mathcal{O}_X}(F,\omega)$. Clearly $E|_U = E$.

Now $cd(X-Y) < n-1$ implies (by definition) that $H^i(X-Y,F) = 0$ for $i \geq n-1$. Hence the local cohomology exact sequence

$$0 = H^{n-1}(X-Y,F) \longrightarrow H^n_Y(X,F) \longrightarrow H^n(X,F) \longrightarrow H^n(X-Y,F) = 0$$

gives an isomorphism

$$H^n_Y(X,F) \xrightarrow{\cong} H^n(X,F) \quad . \tag{i}$$

Since F is locally free in a neighborhood of Y, the proof of Theorem 3.3, Chapter III, applies to F and so we have

$$H^o(\hat{X},\hat{E}) \cong (H^n_Y(X,F))' \tag{ii}$$

where ' denotes dual vector space. On the other hand by Serre duality (applied to F on X), we get that

$$\begin{aligned} H^n(X,F) &\cong (\mathrm{Hom}\,(F,\omega))' \\ &\cong H^o(X, \underline{\mathrm{Hom}}\,(F,\omega))' \\ &\cong H^o(X,E)' \end{aligned} \tag{iii}$$

Thus by (i), (ii) and (iii) above, we have

$$H^o(X,E) \xrightarrow{\cong} H^o(\hat{X},\hat{E}) \ .$$

Since E is torsion free, $H^o(X,E) \longrightarrow H^o(U,E)$ is injective. On the other hand, $H^o(U,E) \longrightarrow H^o(\hat{X},\hat{E})$ is injective because E is locally free on U, and U is connected (in fact irreducible). Therefore we have

$$H^o(X,E) \xrightarrow{\text{inj}} H^o(U,E) \xrightarrow{\text{inj}} H^o(\hat{X},\hat{E})$$

(with the composite $H^o(X,E) \to H^o(\hat{X},\hat{E})$ marked isom)

This implies that all three are isomorphisms. The second isomorphism implies Lef (X,Y).

To prove the second part of (ii), let D be an effective Cartier divisor on X. We claim that $Y \cap D \neq \emptyset$. Indeed, define $\omega_D = \omega_X \otimes \mathcal{O}_D(D)$ where $\omega_X = \Omega^n_{X/k}$. We have an exact sequence

$$0 \longrightarrow \omega_X \longrightarrow \omega_X(D) \longrightarrow \omega_D \longrightarrow 0$$

and hence the cohomology exact sequence

$$H^{n-1}(D,\omega_D) \longrightarrow H^n(X,\omega_X) \longrightarrow H^n(X,\omega_X(D)) .$$

Now by duality $H^n(X,\omega_X(D)) \cong H^0(X,\mathcal{O}_X(-D))' = 0$ and $H^n(X,\omega_X) \cong k$. Hence $H^{n-1}(D,\omega_D) \neq 0$. This shows D cannot be contained in $X-Y$, because $\mathrm{cd}\,(X-Y) < n-1$. Thus $Y \cap D \neq \emptyset$.

(ii) $\Longrightarrow$ (i). By Proposition 3.1 (iii), Chapter III, it is sufficient to prove that $H^i(X-Y,\mathcal{O}_X(m)) = 0$ for all $i \geq n-1$ and all $m \ll 0$.

Suppose L is any invertible sheaf on X. By Lef (X,Y), there exists an open set $U \supseteq Y$ such that $H^0(U,L) \longrightarrow H^0(\hat{X},\hat{L})$ is an isomorphism. Since Y meets every effective Cartier divisor, and $Y \subseteq U$, we have $\mathrm{codim}\,(X-U,X) \geq 2$. Hence we have $H^0(X,L) \xrightarrow{\cong} H^0(U,L) \xrightarrow{\cong} H^0(\hat{X},\hat{L})$. But then by duality (Theorem 3.3, Chapter III) we have

$$H^0(\hat{X},\hat{L}) \cong (H^n_Y(X,M))' ,$$

where $M = \underline{\mathrm{Hom}}\,(L,\omega_X)$. Now the exact sequence

$$\begin{array}{ccccccc}
H^{n-1}(X,M) & \longrightarrow & H^{n-1}(X-Y,M) & \longrightarrow & H^n_Y(X,M) & \longrightarrow & H^n(X,M) \\
\| \wr & & & & \| \wr & & \| \wr \\
H^1(X,L)' & & & & H^o(\hat{X},\hat{L})' & \longrightarrow & H^o(X,L)'
\end{array}$$

gives that $H^{n-1}(X,M) \longrightarrow H^{n-1}(X-Y,M)$ is surjective. Thus if $L = \omega_X(m)$, we have $H^1(X,L) = 0$ for $m >> 0$. But then $M = \mathcal{O}_X(-m)$. Hence $H^{n-1}(X-Y,\mathcal{O}_X(-m)) = 0$ for $m >> 0$. This completes the proof.

Corollary 1.2. Let X be a non-singular projective variety of dimension n. Let $Y \subseteq X$ be a set-theoretic intersection (not necessarily complete intersection) of at most n-1 effective ample divisors. Then Lef (X,Y). (This result applies in particular if Y is a set-theoretic complete intersection and of dimension ≥ 1).

Proof. In this case $X-Y$ is a union of at most n-1 affine open subsets. So by the proof of Theorem 5.1, Chapter III, we get that cd $(X-Y) < n-1$ and hence Lef (X,Y).

Proposition 1.3. Let X be a non-singular complete scheme. Let $Y \subseteq X$ be a complete intersection of r linearly equivalent effective divisors $D_1,\ldots,D_r$ where $r = \operatorname{codim}(Y,X)$. Let $\mathcal{O}_X(1)$ be the invertible sheaf corresponding to the D_i's. Assume that $\dim Y \geq 2$, and $\mathcal{O}_Y(1)$ is ample on Y. Then for any locally free sheaf $\mathfrak{F}$ on $\hat{X}$ (= the formal completion along Y), the sheaf $\mathfrak{F}(m) = \mathfrak{F} \otimes \mathcal{O}_X(m)$ is generated by global sections for $m >> 0$.

Proof. Let $I_1,\ldots,I_r$ be the sheaves of ideals of the effective divisors $D_1,\ldots,D_r$. For each $k = 1,\ldots,r$, let $\mathfrak{F}_k = \mathfrak{F}/(I_1+\ldots+I_k)\mathfrak{F}$. Write $\mathfrak{F}_o = \mathfrak{F}$. Note that $\mathfrak{F}_r$ is a coherent (algebraic) sheaf on Y, in fact it is locally free.

Let $s_i \in H^o(X,\mathcal{O}_X(1))$ be the sections defining the effective divisors D_i, $i = 1,\ldots,r$. Let S be the polynomial ring $k[x_o,\ldots,x_r]$. For each $k = 1,\ldots,r$, let

$$M^k = \sum_{m\geq 0} H^1(\hat{X},\mathfrak{J}_k(m)) .$$

Clearly each M^k is a graded S-module via the homomorphism $S \longrightarrow \sum_{m\geq 0} H^o(X,\mathcal{O}_X(m))$ defined by $x_i \longmapsto s_i$.

We proceed in several steps.

1. For each $i < \dim Y$, $H^i(\hat{X}, \mathfrak{J}_k(m))$ is finite-dimensional for all k and all m.

Indeed, we know that the normal bundle to Y is $\mathcal{O}_Y(1)^{\oplus r}$ which is ample and Γ-ample. Now the assertion follows by the technique of the proof of Theorem 4.1 and the following remark, Chapter III.

2. We will show by descending induction on k, that the graded S-modules M^k are of finite type.

Indeed, for $k = r$ we have $(M^r)_m = H^1(\mathfrak{J}_r(m)) = 0$ for $m \gg 0$ (since $\mathfrak{J}_r$ is a coherent algebraic sheaf on Y and $\mathcal{O}_Y(1)$ is ample on Y), and each $H^1(\mathfrak{J}_r(m))$ is finite-dimensional. Hence M^r is of finite type over S.

Consider the exact sequences

$$0 \longrightarrow I_k\mathfrak{J}_{k-1} \longrightarrow \mathfrak{J}_{k-1} \longrightarrow \mathfrak{J}_k \longrightarrow 0 .$$

By definition, we have $I_k \cong \mathcal{O}_X(-1)$ for all $k = 1,\ldots,r$. Twisting by $\mathcal{O}_X(m)$, we get exact sequences

$$0 \longrightarrow \mathcal{J}_{k-1}(m-1) \longrightarrow \mathcal{J}_{k-1}(m) \longrightarrow \mathcal{J}_k(m) \longrightarrow 0$$

and hence the cohomology exact sequences

$$(*): \; H^0(\mathcal{J}_{k-1}(m)) \longrightarrow H^0(\mathcal{J}_k(m)) \longrightarrow$$

$$H^1(\mathcal{J}_{k-1}(m-1)) \xrightarrow{x_k} H^1(\mathcal{J}_{k-1}(m)) \xrightarrow{\alpha} H^1(\mathcal{J}_k(m)) .$$

Suppose inductively, we have proved that M^k is of finite type for a given k. To prove that M^{k-1} is of finite type: The cohomology sequence above gives an exact sequence of S-modules

$$M^{k-1} \xrightarrow{x_k} M^{k-1} \xrightarrow{\alpha} M^k.$$

Since M^k is of finite type (by induction), we get that $M^{k-1}/x_k M^{k-1}$ is of finite type. But then the following lemma gives that M^{k-1} is of finite type. This proves (2).

<u>Lemma 1.4</u>. Let $M = \sum_{m\geq 0} M_m$ be a graded module over a graded noetherian ring S, with $S_0 = k$. Let $x \in S_1$. Then M/xM is of finite type $\Longrightarrow$ M is of finite type.

<u>Proof</u>. Indeed, suppose $z_i \in M_{m_i}$, $i = 1,\ldots,r$, are such that their residues generate M/xM. Then given $z \in M_m$, we can write $z = y' + xz'$ where $y' \in \sum_{i=1}^{r} Sz_i$ and $z' \in M_{m-1}$. By repeating this process (for z' and so on), we get an expression of the form $z = y + x^{m+1}z^*$ with $y \in \sum_i Sz_i$ and $z^* \in M_{-1} = 0$. Thus M itself is generated by the z_i.

3. We can redefine $x_1,\dots,x_r$ by a linear change of variables (if necessary) such that each $M^{k-1} \xrightarrow{x_k} M^{k-1}$ is injective in sufficiently large degrees.

We do this by ascending induction on k. By (2), each M^k is of finite type. Writing $M = M^o$, and looking at $M_{\geq m_o}$ (for some m_o) instead of M, we may assume that the ideal $\mathfrak{M} = (x_1,\dots,x_r)$ is not associated to M. Taking any x_1' which is outside every prime ideal associated to M, we get that $M \xrightarrow{x_1'} M$ is injective in large degrees. Note that $x_1' \in S_1$. Now looking at $M^1 = M/x_1'M$, we can find an $x_2' \in S_1$ which is linearly independent of x_1', and such that $M^1 \xrightarrow{x_2'} M^1$ is injective in degrees $\gg 0$. Proceeding thus we get $x_1',\dots,x_r' \in S_1$ such that each $M^{k-1} \xrightarrow{x_k'} M^{k-1}$ is injective in sufficiently large degrees.

Suppose D_i' is the effective divisor defined by the image of x_i' in $H^o(X,\mathcal{O}_X(1))$ and I_i' is the corresponding sheaf of ideals. Then define $\mathfrak{F}_k' = \mathfrak{F}/(I_1'+\dots+I_k')\mathfrak{F}$. Now replace $D_i, x_i, \mathfrak{F}_k$ by $D_i', x_i', \mathfrak{F}_k'$ throughout. The hypotheses and all of the proof so far still hold, so we may assume that each $M^{k-1} \xrightarrow{x_k} M^{k-1}$ is injective in sufficiently large degrees.

4. We will show by descending induction on k, that each $\mathfrak{F}_k(m)$ is generated by global sections for $m \gg 0$.

Indeed, for $k = r$, $\mathfrak{F}_r$ is a coherent algebraic sheaf on Y, and $\mathcal{O}_Y(1)$ is ample on Y, and so $\mathfrak{F}_r(m)$ is generated by global sections for $m \gg 0$.

Suppose inductively, we have proved (4) for a given k. To prove for $k-1$: Consider the exact sequences (see (*) in (2) above)

$$\begin{array}{ccccccc} H^0(\mathfrak{F}_{k-1}(m)) & \longrightarrow & H^0(\mathfrak{F}_k(m)) & \longrightarrow & H^1(\mathfrak{F}_{k-1}(m-1)) & \xrightarrow{x_k} & H^1(\mathfrak{F}_{k-1}(m)) \\ & & & & \| & & \| \\ & & & & (M^{k-1})_{m-1} & \xrightarrow{x_k} & (M^{k-1})_m \end{array}.$$

By (3), the last arrow is injective for $m \gg 0$, and hence the first arrow is surjective. But $\mathfrak{F}_k(m)$ is generated by global sections (by induction), and hence $\mathfrak{F}_{k-1}(m)$ is generated by global sections (by Nakayama) for $m \gg 0$. This proves (4).

This completes the proof of the proposition too.

<u>Theorem 1.5</u>. Let X be a non-singular subvariety of $\mathbb{P} = \mathbb{P}_k^N$, and let $Y \subseteq X$ be a closed subscheme of codimension r, which is a complete intersection (i.e., $Y = X \cap H_1 \cap \ldots \cap H_r$ where the H_i are hypersurfaces in $\mathbb{P}$). Assume that $\dim Y \geq 2$. Then we have Leff (X,Y).

<u>Proof</u>. Since Y is a complete intersection, by Corollary 1.2, we have Lef (X,Y). Replacing each H_i by a suitable multiple, we may assume that the H_i have the same degree d. Then replacing $\mathbb{P}$ by its d-uple imbedding, we may assume that the H_i are hyperplanes. (This may change the scheme structure on Y, but our conclusion does not depend on this).

Let $\hat{X}$ be the formal completion of X along Y, and let $\mathfrak{F}$ be a locally free sheaf on $\hat{X}$. By Proposition 1.3, $\mathfrak{F}(m)$ is generated by global sections for $m \gg 0$. Therefore we have an exact sequence

$$\mathcal{O}_{\hat{X}}(-m)^{\oplus n_2} \xrightarrow{\tilde{\varphi}} \mathcal{O}_{\hat{X}}(-n)^{\oplus n_1} \longrightarrow \mathfrak{J} \longrightarrow 0$$

for suitable integers m,n and n_i, $i = 1,2$. Now consider the sheaf on X

$$E = \underline{\text{Hom}}\ (\mathcal{O}_X(-m)^{\oplus n_2}, \mathcal{O}_X(-n)^{\oplus n_1}) .$$

By Lef (X,Y), there exists an open set $U \supseteq Y$ such that the natural map

$$H^0(U,E) \xrightarrow{\cong} H^0(\hat{X},\hat{E})$$

is an isomorphism, and so $\tilde{\varphi}$ comes from a homomorphism on U

$$\mathcal{O}_U(-m)^{\oplus n_2} \xrightarrow{\varphi} \mathcal{O}_U(-n)^{\oplus n_1} .$$

Let $F = \text{coker}\ \varphi$. Then clearly F is coherent, and is locally free on an open set U' with $Y \subseteq U' \subseteq U$. By construction we have $\hat{F} \cong \mathfrak{J}$. Thus we have Leff (X,Y).

Exercise 1.6. Show that in the above theorem, $\dim Y \geq 2$ is necessary. In fact, show that the natural embedding of $Y = \mathbb{P}^1_k$ in $X = \mathbb{P}^2_k$ does not satisfy Leff (X,Y).

Exercise 1.7. Let X be a complete non-singular surface and let $Y \subsetneq X$ be a non-singular curve. Then show that Lef $(X,Y) \Longleftrightarrow (Y^2) > 0$.

Exercise 1.8. Let X be an integral scheme, and let Y be a closed subscheme satisfying Lef (X,Y). Then show that Y is connected.

Problem 1.9. Give necessary and sufficient conditions for a subvariety $Y \subseteq \mathbb{P}^n_k$ to satisfy the effective Lefschetz condition $\mathrm{Leff}(\mathbb{P}^n_k, Y)$.

§2. Application to the fundamental group.

For any scheme or formal scheme X we denote by Et (X) the category of finite étale covers of X (see Grothendieck [SGA 1] exposé I, for the theory of étale morphisms). The algebraic fundamental group of X is constructed from the category Et (X) by a well-known procedure (see Grothendieck [SGA 1] exposé V, or Murre [1]).

Theorem 2.1. Let X be a complete non-singular variety and let Y be a closed subscheme. Assume

(i) Leff (X,Y), and

(ii) Y meets every effective divisor on X.

Then the natural restriction map

$$\text{Et}(X) \longrightarrow \text{Et}(Y)$$

is an equivalence of categories, and hence $\pi_1(Y) \longrightarrow \pi_1(X)$ is an isomorphism.

Proof. Let U be any open neighborhood of Y in X, and let $\hat{X}$ be the formal completion of X along Y. Then we have natural restriction functors

$$\text{Et}(X) \longrightarrow \text{Et}(U) \longrightarrow \text{Et}(\hat{X}) \longrightarrow \text{Et}(Y).$$

We will show that each of these is an equivalence of categories.

1. Et $(X) \longrightarrow$ Et (U) is an equivalence of categories.

Since Y meets every effective divisor on X, we have codim $(X-U, X) \geq 2$. Now every étale covering of U extends to a (possibly branched) covering of X in a unique way. But by the purity of the branch locus (see Zariski [5] or Nagata [4], Theorem (41.1), p. 158), it must be étale.

2. Et (U) $\longrightarrow$ Et ($\hat{X}$) is an equivalence of categories.

To give an étale cover of U or $\hat{X}$ is the same as giving a locally free sheaf of commutative algebras with unit which is étale at each point. A homomorphism of algebras is in particular a homomorphism of sheaves, so already the weak Lefschetz condition Lef (X,Y) implies that the functor is fully faithful.

For the existence, let $\mathcal{A}$ be a locally free sheaf of commutative algebras with unit on $\hat{X}$. By Leff (X,Y), we can find a locally free sheaf A on a suitable neighborhood U of Y such that $\hat{A} \cong \mathcal{A}$. The algebra structure on $\mathcal{A}$ is given by a homomorphism $\mu \in \operatorname{Hom}(\mathcal{A} \times \mathcal{A}, \mathcal{A})$. So by Lef (X,Y), this can be extended to a certain neighborhood. The properties of commutativity and associativity of the multiplication are expressed by certain identities which hold on $\hat{X}$, and hence in a neighborhood. Similarly the unit can be extended. Thus, after shrinking U suitably, we find that A is a sheaf of commutative algebras with unit. Finally the property of being étale holds on an open set.

Thus we see that any étale cover of X extends to a suitable neighborhood of Y. By virtue of 1 above, it does not matter which neighborhood. Thus Et (U) $\longrightarrow$ Et ($\hat{X}$) is an equivalence for any U.

3. Et $(\hat{X}) \longrightarrow$ Et (Y) is an equivalence of categories.

This follows from the general theory of étale morphisms (see Grothendieck [SGA 1], exposé I, Corollaire 8.4, p. 15).

Corollary 2.2. Let X be a non-singular projective variety, let Y be a subscheme of dimension ≥ 2 which is a complete intersection in X. Then

$$\pi_1(Y) \longrightarrow \pi_1(X)$$

is an isomorphism.

Proof. Indeed, in that case we have Leff (X,Y) by Theorem 1.5, and clearly Y meets every effective divisor on X.

§3. Application to the Picard group.

Theorem 3.1. Let X be a complete non-singular variety, and let Y be a closed subscheme. Assume

(i) Leff (X,Y),

(ii) Y meets every effective divisor on X, and

(iii) $H^i(Y, I^n/I^{n+1}) = 0$ for $i = 1,2$, and all $n \geq 1$, where I is the sheaf of ideals of Y.

Then the natural map

$$\text{Pic } X \longrightarrow \text{Pic } Y$$

is an isomorphism.

Proof. Let U be any open neighborhood of Y and let $\hat{X}$ denote the completion of X along Y. As in the previous section, we consider the natural restriction homomorphisms

$$\text{Pic } X \longrightarrow \text{Pic } U \longrightarrow \text{Pic } \hat{X} \longrightarrow \text{Pic } Y ,$$

and we will show that they are all isomorphisms.

1. Pic $X \longrightarrow$ Pic U is an isomorphism.

Since Y meets every effective divisor on X, we have codim $(X-U, X) \geq 2$. Now since X is non-singular, all its local rings are UFD, so every invertible sheaf on U extends uniquely to an invertible sheaf on X.

2. Pic $U \longrightarrow$ Pic $\hat{X}$ is an isomorphism.

Indeed, by Leff (X,Y), every invertible sheaf on $\hat{X}$ extends uniquely to an invertible sheaf on a suitable neighborhood of Y. Combining with 1, we find Pic $U \longrightarrow$ Pic $\hat{X}$ is an isomorphism.

3. Pic $\hat{X} \longrightarrow$ Pic Y is an isomorphism.

For this step we factorize further. Let I be the sheaf of ideals of Y, and for each $n \geq 1$ let Y_n be the subscheme defined by I^n. Then we have natural maps

$$\mathrm{Pic}\ \hat{X} \longrightarrow \varprojlim \mathrm{Pic}\ Y_n \longrightarrow \cdots \longrightarrow \mathrm{Pic}\ Y_{n+1} \longrightarrow \mathrm{Pic}\ Y_n \longrightarrow \cdots \longrightarrow \mathrm{Pic}\ Y.$$

We will show that these are all isomorphisms. The first isomorphism is an easy exercise (see for example Hartshorne [4], Lemma 8.2, p. 446). To compare Pic Y_{n+1} and Pic Y_n, we use the exact sequence

$$0 \longrightarrow I^n/I^{n+1} \longrightarrow \mathcal{O}^*_{Y_{n+1}} \longrightarrow \mathcal{O}^*_{Y_n} \longrightarrow 0 ,$$

where the first map is the "exponential" which sends x to $1 + x$, and where $*$ denotes the multiplicative group of units in the given ring. This gives a long exact sequence of cohomology

$$\cdots \longrightarrow H^1(Y, I^n/I^{n+1}) \longrightarrow H^1(Y_{n+1}, \mathcal{O}^*_{Y_{n+1}}) \longrightarrow H^1(Y_n, \mathcal{O}^*_{Y_n})$$
$$\longrightarrow H^2(Y, I^n/I^{n+1}) \longrightarrow \cdots .$$

Now our hypothesis (iii) says that the two outside groups are zero. Hence the middle ones, which are just Pic Y_{n+1} and Pic Y_n, are isomorphic.

Since all the groups Pic Y_n are isomorphic, their inverse limit is also isomorphic to them, and we are done.

<u>Corollary 3.2</u>. Let Y be a subscheme of dimension ≥ 3 which is a (strict) complete intersection in $\mathbb{P}^n_k$. Then Pic $Y \cong \mathbb{Z}$, and it is generated by the class of $\mathcal{O}_Y(1)$.

<u>Proof</u>. Since Y is a complete intersection, we have Leff (X,Y) by Theorem 1.5, and Y meets every divisor on $\mathbb{P}_k^n$. Furthermore, if Y is an intersection of hypersurfaces of degree $d_1,\ldots,d_r$ then I/I^2 is isomorphic to $\mathcal{O}_Y(-d_1) \oplus \cdots \oplus \mathcal{O}_Y(-d_r)$. Hence I^n/I^{n+1} is a direct sum of sheaves of the form $\mathcal{O}_Y(m_i)$ for suitable integers $m_i < 0$. But Serre ([FAC], §78, Proposition 5 (b), p. 273) has shown that $H^i(Y,\mathcal{O}_Y(m)) = 0$ for any complete intersection Y, for all $0 < i < \dim Y$, and all $m \in \mathbb{Z}$. Thus since $\dim Y \geq 3$, the hypotheses of the theorem are satisfied.

<u>Corollary 3.3</u>. Let X be a non-singular projective variety over $\mathbb{C}$, and let Y be a non-singular subvariety of dimension ≥ 3, which is a (strict) complete intersection in X. Then the natural map

$$\mathrm{Pic}\, X \longrightarrow \mathrm{Pic}\, Y$$

is an isomorphism.

<u>Proof</u>. As in the proof of the previous Corollary, we need only check that $H^i(Y, \mathcal{O}_Y(m)) = 0$ for $i = 1,2$, and all $m < 0$. But this follows from Kodaira's vanishing theorem (see Kodaira [1], also Chapter VI, §1 (i), below).

<u>Exercise 3.4</u>. A subvariety $Y \subseteq \mathbb{P} = \mathbb{P}^n_k$ is called <u>projectively normal</u> if it is normal and the natural maps $H^0(\mathbb{P}, \mathcal{O}_{\mathbb{P}}(m)) \longrightarrow H^0(Y, \mathcal{O}_Y(m))$ are isomorphisms for all m. Show that a non-singular complete intersection is projectively normal.

<u>Exercise 3.5</u>. Let Y be a non-singular subvariety of $\mathbb{P} = \mathbb{P}^n_k$. Show that the following conditions are equivalent:

(i) Pic $Y \cong \mathbb{Z}$, generated by the class of $\mathcal{O}_Y(1)$, and Y is projectively normal,

(ii) for every effective Cartier divisor $Z \subseteq Y$, there is a hypersurface $H \subseteq \mathbb{P}$ with $Z = Y \cap H$, as schemes,

(iii) the homogeneous coordinate ring $A(Y) = k[x_o, \ldots, x_n]/I_Y$ is a unique factorization domain.

§4. Historical note.

There are many theorems comparing cohomological invariants of an algebraic variety with its hyperplane section. We will give a brief survey of such theorems in this section.

(i) Theorems on Pic.

Lefschetz ([1], p. 359) claims the first complete proof of "Noether's theorem". His statement is this.

Theorem. Let Y be a non-singular s-dimensional variety, $s \geq 3$, which is a (strict) complete intersection in $\mathbb{P}^n_{\mathbb{C}}$. Then Y contains only hypersurfaces which are themselves complete intersections. The same is true for a sufficiently general surface Y, which is a complete intersection in $\mathbb{P}^n$, $n \geq 4$, and which is not contained in any $\mathbb{P}^{n-1}$. Finally, the same result holds for a sufficiently general surface Y of degree $d \geq 4$ in $\mathbb{P}^3$.

The earliest reference to this result seems to be M. Noether ([1], pp. 58, 64) who asserts that a general surface of degree $d \geq 4$ in $\mathbb{P}^3$ contains only curves which are complete intersections. Severi [1] extended this result to non-singular hypersurfaces in $\mathbb{P}^n$, $n \geq 4$. And Fano [1] made essentially the same assertion as in Lefschetz' theorem.

More recent work includes a difficult paper by Andreotti and Salmon [1]. Their proof, in the case $\dim Y \geq 3$, is valid also in characteristic p. Another proof is given by Franchetta [1].

A completely modern proof, for the case $\dim Y \geq 3$, is given by Grothendieck [SGA 2]. It is this proof which we have followed in this chapter. However, these methods have not yet been successfully applied to the proof of Noether's theorem, for surfaces in $\mathbb{P}^3$. There is a new proof of Noether's theorem (over $\mathbb{C}$) by Moisezon [3] and another by Deligne (unpublished).

(ii) Theorems on π_1.

Picard and Simart ([1], p. 85 ff) prove that any non-singular surface in $\mathbb{P}^3$ is simply connected. Zariski [1] studies the fundamental group of projective space minus a hypersurface. His theorem is this

Theorem. Let Y be a hypersurface in $\mathbb{P}^n_{\mathbb{C}}$, $n \geq 3$. Let H be a hyperplane in $\mathbb{P}^n$, in general position. Then the inclusion

$$H - H \cap X \longrightarrow \mathbb{P}^n - X$$

induces an isomorphism of the fundamental groups.

Abhyankar [1] and Grothendieck [SGA 1] have defined a purely algebraic fundamental group using finite unramified coverings. This algebraic fundamental group can be compared with the usual topological one over $\mathbb{C}$ by means of the "generalized Riemann existence theorem" (see Artin and Grothendieck [SGAA], exposé XI, 4.3).

The "Lefschetz theorem" for π_1 was proved in its algebraic form by Grothendieck [SGA 2]. A proof of the topological form is given by Bott [1].

It would be interesting to give a proof of Zariski's theorem using the algebraic methods developed by Grothendieck. More generally, it seems reasonable to compare the fundamental group of any quasi-projective variety with its general hyperplane section.

(iii) Theorems on integral homology.

Again, the basic theorem was proved by Lefschetz [1], p. 331.

Theorem. Let X be a non-singular projective algebraic variety over $\mathbb{C}$, and let Y be a non-singular hyperplane section of X. Then the natural maps

$$H_i(Y,\mathbb{Z}) \longrightarrow H_i(X,\mathbb{Z})$$

are isomorphisms for $i < \dim Y$, and surjective for $i = \dim Y$.

Lefschetz devotes only two pages to the proof. However, Wallace [1] has expanded Lefschetz' proof to include all necessary topological details.

More recent proofs of this theorem using Morse theory are given by Andreotti and Frankel [1] and Bott [1]. A sheaf-theoretic proof was given by Fáry [1], who also determines the complete structure of the integral homology of a general complete intersection in $\mathbb{P}^n$.

(iv) Theorems on cohomology.

Of course one can obtain theorems on integral cohomology by applying the universal coefficient theorem to those on homology.

Kodaira and Spencer [1], and Hirzebruch [1] have proved Lefschetz-type theorems for the cohomology groups $H^q(X,\Omega^p)$.

Artin and Grothendieck ([SGAA], exposé XIV, Corollaire 3.2, p. 16) have proved an "affine Lefschetz theorem" for étale cohomology, which says that the étale cohomology of a non-singular affine scheme U of dimension n is zero in degrees > n. On the other hand, they show (exposé X, Corollaire 4.3, p. 14) that the étale cohomological dimension (écd) of any scheme of dimension n is at most 2n. This suggests the following

Problem 4.1. Let X be a scheme of dimension n over an algebraically closed field k. Is it true that écd X $\leq$ cd X + n? (See Proposition 7.2, Chapter III for an analogue using algebraic De Rham cohomology).

CHAPTER V

FORMAL-RATIONAL FUNCTIONS ALONG A SUBVARIETY

In this chapter we will consider the ring $K(\hat{X})$ of formal-rational functions on a variety X along a subvariety Y. They are functions which are locally quotients of regular functions on the formal completion $\hat{X}$ of X along Y. They can be considered as a formal analogue to meromorphic functions in a neighborhood of Y, for complex-analytic spaces. In good cases $K(\hat{X})$ is a field.

This field was first studied by Hironaka [6] and Hironaka and Matsumura [1]. They proved that in certain cases (for example if Y is an ample divisor on X, or Y is a connected positive-dimensional subvariety of $X = \mathbb{P}^n_k$) the field $K(\hat{X})$ is just equal to the function field K(X) of X. (In this case we say Y is G3 in X.) This has significance for birational geometry. If Y is G3 in X, then X is birationally uniquely determined by $\hat{X}$. Furthermore, if $Y \longrightarrow X'$ is any other embedding of Y in a variety X', with $\hat{X} \cong \hat{X}'$, then there is a unique birational map $X \longrightarrow X'$, which is a morphism in a neighborhood of Y, and which induces the isomorphism $\hat{X} \cong \hat{X}'$. Finally, if $K(\hat{X})$ is only a finite algebraic extension of K(X) (in which case we say Y is G2 in X) then there is an "étale neighborhood" of Y in which Y is G3 (see Exercise 4.9, below).

In this chapter we will prove these theorems of Hironaka and Matsumura. In the course of developing this theory, we will find there are close connections between the properties G2, G3, the

Lefschetz conditions of Grothendieck, and the cohomological dimension of the complement of the subvariety.

Our first aim is to show that a complete intersection Y in a projective variety X is G3. If $\dim Y \geq 2$, this follows from the effective Lefschetz condition (Corollary 1.4). If $\dim Y = 1$, we must use a different approach. The key result here is Theorem 1.6, which is a special case of a theorem of Hartshorne [CDAV] which says that any locally complete intersection with ample normal bundle is G2. Then using the fact that G3 $\Longleftrightarrow$ G2 + Lef (Proposition 2.1), we deduce that any complete intersection curve is G3.

From here we can prove that any connected positive-dimensional subvariety Y of projective space $\mathbb{P}^n_k$ is G3, using Hironaka's method of projection (Theorem 3.1). This in turn implies that $cd\,(\mathbb{P}^n - Y) < n-1$ (Theorem 3.2).

At the end of §3 we give some new examples of non-algebraizable formal schemes.

Two topics we have not included, for lack of time, are the behavior under a proper morphism, and the application to abelian varieties. Hironaka and Matsumura ([1], Theorem 2.7, p. 61) prove the following: let $f: X' \longrightarrow X$ be a proper surjective morphism of non-singular varieties, let Y be a closed subset of X, and let $Y' = f^{-1}(Y)$. Then Y is G2 (resp. G3) in X $\Longleftrightarrow$ Y' is G2 (resp. G3) in X'. They also prove, in §4 of their paper, that a subvariety Y of an abelian variety A is G2 $\Longleftrightarrow$ Y generates A.

Furthermore, Y is G3 $\Longleftrightarrow$ it generates A, and the natural map Alb Y $\longrightarrow$ A has connected fibres. Speiser [1] uses this result to show that in the latter case cd (A-Y) $<$ n-1, where n = dim A.

For analogous theorems on complex analytic spaces, and the relations between the algebraic and the analytic case, see Chapter VI, §1 (vii), and §2, Exercises 2.9, 2.10.

§1. The ring of formal-rational functions.

In this section we introduce the ring of formal-rational functions along a subvariety, and define the conditions G1, G2, and G3 of Hironaka and Matsumura [1]. We give some elementary properties and examples of these notions. The main result of this section (Theorem 1.6, below) gives a criterion for a curve to be G2 in an ambient variety.

Let X be a scheme, and let $Y \subseteq X$ be a closed subscheme defined by the sheaf of ideals I_Y. Let ^ denote the formal completion along Y. Let $\mathcal{K}_X$ be the sheaf of total quotient rings of $\mathcal{O}_{\hat{X}}$.

The ring $K(\hat{X}) = H^0(\hat{X}, \mathcal{K}_{\hat{X}})$ <u>is called the ring of formal-rational functions along</u> Y, and the subring $H^0(\hat{X}, \mathcal{O}_{\hat{X}})$ is called <u>the ring of formal-regular functions along</u> Y.

<u>Remark</u>. If X is a non-singular variety and Y is connected, then $K(\hat{X})$ is a field. This is a consequence of the following fact.

We say that a (noetherian) ring A is <u>regular</u> if the localizations at all its prime ideals are regular local rings. Let A be a regular ring, and let $J \subseteq A$ be an ideal. Let $\hat{A}$ denote the separated completion of A for the J-adic topology. Then by [EGA, 0_{IV}, 17.3.8.1], $\hat{A}$ is a regular ring.

Now if X is a non-singular variety, for every affine open subset $U = \mathrm{Spec}\ (A)$ in X, A is a regular domain. Therefore for every connected (formal) affine open subset $\hat{U} \subseteq \hat{X}$, $H^0(\hat{U}, \mathcal{O}_{\hat{X}})$ is a regular domain. So $H^0(\hat{U}, \mathcal{K}_{\hat{X}})$ is a field. Now Y connected implies that $K(\hat{X})$ is a field.

We make the following definitions due to Hironaka and Matsumura [1, p. 64]. We say that

1) Y <u>is</u> G1 <u>in</u> X if the natural map $H^o(X, \mathcal{O}_X) \longrightarrow H^o(\hat{X}, \mathcal{O}_{\hat{X}})$ is an isomorphism;

2) Y <u>is</u> G2 <u>in</u> X if $K(\hat{X})$ is a finite module over $K(X)$ via the natural map $K(X) \longrightarrow K(\hat{X})$;

3) Y <u>is</u> G3 <u>in</u> X if $K(X) \cong K(\hat{X})$.

<u>Remarks</u>. 1. It is obvious that G3 $\Longrightarrow$ G2.

2. Suppose X is an algebraic variety over k such that $H^o(X, \mathcal{O}_X) = k$ (for example X a complete variety). Then $K(X)$ is a field and $K(\hat{X})$ is a finite direct sum of fields. If Y is G1 or G3 in X, then Y must be connected.

<u>Proposition 1.1</u>. Let X be a complete non-singular scheme, and let $Y \subseteq X$ be a connected closed subscheme. Then Y is G2 in $X \Longrightarrow Y$ is G1 in X.

<u>Proof</u>. We first prove the following

<u>Lemma 1.2</u>. Let X be a complete scheme, and let Y be a connected closed subscheme. Then the ring $A = H^o(\hat{X}, \mathcal{O}_{\hat{X}})$ is a complete local ring with residue field k.

<u>Proof</u>. We may assume that Y is reduced. Let I_Y be the sheaf of ideals defining Y. Let $\mathfrak{M} = H^o(\hat{X}, \hat{I}_Y)$. We have an exact sequence

$$0 \longrightarrow \hat{I}_Y \longrightarrow \mathcal{O}_{\hat{X}} \longrightarrow \mathcal{O}_Y \longrightarrow 0$$

and hence the cohomology exact sequence

$$\begin{array}{ccccccc} 0 \longrightarrow & H^0(\hat{X},\hat{I}_Y) & \longrightarrow & H^0(\hat{X},\mathcal{O}_{\hat{X}}) & \longrightarrow & H^0(Y,\mathcal{O}_Y) & \\ & \| & & \| & & \| & \\ 0 \longrightarrow & \mathfrak{M} & \longrightarrow & A & \longrightarrow & k & . \end{array}$$

Thus $\mathfrak{M}$ is a maximal ideal. Since $\mathcal{O}_{\hat{X}}$ is complete with respect to the ideal $\hat{I}_Y$, we find that for $x \in \mathfrak{M}$, $1 - x + x^2 - \cdots \in A$ and so $1 + x$ is a unit in A. Hence $\mathfrak{M}$ is the ideal of non-units, so A is a local ring.

Let $\mathfrak{A}_n = H^0(\hat{X},\hat{I}_Y^n)$ and let $B_n = H^0(\hat{X},\mathcal{O}_{\hat{X}}/\hat{I}_Y^n)$. Then $\mathfrak{M}^n \subseteq \mathfrak{A}_n$, and we have exact sequences

$$0 \longrightarrow \mathfrak{A}_n \longrightarrow A \longrightarrow B_n .$$

On the other hand, we have

$$A = \varprojlim_n B_n .$$

Therefore

$$\begin{aligned} A &= \varprojlim_n (\text{Image of } A \text{ in } B_n) \\ &= \varprojlim_n (A/\mathfrak{A}_n) . \end{aligned}$$

Thus A is complete for the $(\mathfrak{A}_n)$-topology. Since $\mathfrak{M}^n \subseteq \mathfrak{A}_n$, A is also complete for the $\mathfrak{M}$-adic topology, i.e., A is a complete local ring.

Remark. We do not know if A is noetherian. This question was raised by Zariski [2].

Proof of Proposition. By G2, $K(\hat{X})$ is a finite algebraic extension field of $K(X)$. In particular, tr. deg. $K(\hat{X})/k$ = tr. deg. $K(X)/k < \infty$. Let $A = H^0(\hat{X},\mathcal{O}_{\hat{X}})$. To prove that Y is G1 in X, we must prove that $A = k$.

By the lemma, A is a complete local ring with residue field k. If $A \neq k$, take any $x \in \mathfrak{m}$, $x \neq 0$. Then $k[[x]] \subseteq A$, and so $k((x)) \subseteq K(\hat{X})$. This shows that

$$\text{tr. deg. } K(\hat{X})/k = \infty$$

which is a contradiction. This completes the proof.

Proposition 1.3. Let X be a non-singular variety, and let $Y \subseteq X$ be a closed subscheme. Then Leff $(X,Y) \Longrightarrow Y$ is G3 in X.

Proof. Given $\xi \in K(\hat{X})$, let $\mathfrak{P} = \mathfrak{P}_\xi$ be the pole sheaf of ξ, i.e., $\mathfrak{P} \subseteq \mathcal{K}_{\hat{X}}$ is the subsheaf defined locally as $\{g \in \mathcal{K}_{\hat{X}} \mid g\xi \in \mathcal{O}_{\hat{X}}\}$. Let $\mathfrak{Q} = \xi\mathfrak{P}$. Since X is non-singular, $\mathfrak{P}$ and $\mathfrak{Q}$ are invertible sheaves on $\hat{X}$. Note that $\xi \in \text{Hom}(\mathfrak{P},\mathfrak{Q})$.

By Leff (X,Y), there exist invertible sheaves P and Q on a neighborhood $U \supseteq Y$ such that $\hat{P} \cong \mathfrak{P}$ and $\hat{Q} \cong \mathfrak{Q}$. Also there exists a $\zeta \in \text{Hom}(P,Q)$ such that ξ is given by ζ. But ζ defines an element of $K(X)$ whose image in $K(\hat{X})$ is ξ. Thus $K(X) \xrightarrow{\cong} K(\hat{X})$ i.e., Y is G3 in X.

Corollary 1.4. Let X be a non-singular projective variety, and let $Y \subseteq X$ be a complete intersection of dimension ≥ 2. Then Y is G3 in X.

Proof. This follows from the Proposition and Theorem 1.5, Chapter IV.

Remark. We will see later (Corollary 2.3, below) that this result holds also for Y of dimension 1.

Corollary 1.5. (Hironaka [6], Theorem IV*, p. 602). Let X be a non-singular complete variety of dimension ≥ 3, and let Y be an effective Cartier divisor with ample normal bundle. Then Y is G3 in X.

Proof. Since the question is local around Y, we may assume that Y is an ample divisor (by Theorem 4.2, Chapter III). Then by Theorem 1.5, Chapter IV, we have Leff (X,Y), and hence Y is G3 in X.

Remark. We will see later (Exercise 4.5, below) that the same result holds also for curves on surfaces. In general, a subvariety with ample normal bundle is G2 but need not be G3. (See Theorem 1.6 and the following remark and example, below).

Now we come to the main theorem of this section.

Let S be a complete irreducible curve, and let L be an invertible sheaf on S. Then we define the degree of L, written deg L, to be the degree of the induced invertible sheaf f^*L on $\tilde{S}_{red}$ where $\sim$ denotes the normalization and $f\colon \tilde{S}_{red} \longrightarrow S$ is the canonical (finite surjective) morphism. Thus for example L is ample on S if and only if deg $L > 0$ (Exercise 3.2 and Proposition 4.4, Chapter I).

Theorem 1.6. Let X be a non-singular complete scheme of dimension n. Let $Y \subseteq X$ be an irreducible curve (not necessarily reduced) such that

i) Y is locally a complete intersection, and

ii) $N_{Y/X} \cong (\mathcal{O}_Y(1))^{\oplus n-1}$ where $\mathcal{O}_Y(1)$ is an ample invertible sheaf on Y.

Then Y is G2 in X.

Proof. We need the following lemmas.

Lemma 1.7. Let Y be a complete irreducible curve. Then there exist integers $a,b,c \in \mathbb{Z}$ such that for all invertible sheaves L on Y, we have

1) $\deg L < a \Longrightarrow H^0(Y,L) = 0$, and

2) $\deg L \geq a \Longrightarrow \dim H^0(Y,L) \leq b \deg L + c$.

Proof. Exercise.

Lemma 1.8. With the hypothesis of the above theorem, there exists a numerical polynomial $P(z) \in \mathbb{Q}[z]$ of degree n, and a constant $a \in \mathbb{Z}$ such that for all invertible sheaves $\mathcal{L}$ on $\hat{X}$, if we define $\deg \mathcal{L} = \deg (\mathcal{L} \otimes \mathcal{O}_Y)$, we have

1) $\deg \mathcal{L} < a \Longrightarrow H^0(\hat{X},\mathcal{L}) = 0$, and

2) $\deg \mathcal{L} \geq a \Longrightarrow \dim H^0(\hat{X},\mathcal{L}) \leq P(\deg \mathcal{L})$.

Proof. Let I_Y be the sheaf of ideals defining Y. Let Y_r be the subscheme defined by I_Y^r, and let $L_r = \mathcal{L} \otimes \mathcal{O}_{Y_r}$. Write $Y = Y_1$ and $L = L_1$. We have $H^0(\hat{X},\mathcal{L}) = \varprojlim_r H^0(Y_r,L_r)$.

We have the exact sequences

$$0 \longrightarrow H^0(L \otimes I_Y^r/I_Y^{r+1}) \longrightarrow H^0(L_{r+1}) \longrightarrow H^0(L_r) \longrightarrow \cdots .$$

Since $(I_Y/I_Y^2)^\vee = N_{Y/X} = (\mathcal{O}_Y(1))^{\oplus n-1}$, we have $I_Y/I_Y^2 = (\mathcal{O}_Y(-1))^{\oplus n-1}$.

Also, since Y has dimension 1, we have $H^o(L \otimes I_Y^r/I_Y^{r+1}) = 0$ for $r >> 0$. Thus $H^o(L_{r+1}) \longrightarrow H^o(L_r)$ is injective for $r >> 0$.

But
$$I_Y^r/I_Y^{r+1} = S^r(I_Y/I_Y^2)$$
$$= S^r((\mathcal{O}_Y(-1)^{\oplus n-1})$$
$$= (\mathcal{O}_Y(-r))^{\oplus N}$$

where $N = \binom{r+n-2}{n-2}$. Hence it follows that

$$\dim H^o(\hat{X},\mathcal{L}) \leq \sum_{r=0}^{\infty} \dim H^o(Y, L \otimes (\mathcal{O}_Y(-r))^{\oplus N})$$
$$= (\text{finite sum}).$$

Now suppose $\deg \mathcal{O}_Y(1) = d$ and $\deg L = e$. We have $\deg (L\otimes\mathcal{O}_Y(-r)) = e - dr$. Let $a,b,c \in \mathbb{Z}$ be as given by Lemma 1.7. If $e - dr < a$, we have $H^o(Y,L\otimes\mathcal{O}_Y(-r)) = 0$ by Lemma 1.7. Thus if $e = \deg \mathcal{L} < a$, we have $H^o(\hat{X},\mathcal{L}) = 0$, which proves (1). On the other hand, if $e - dr \geq a$, we have

$$\dim H^o(Y,L(-r)) \leq b \deg (L(-r)) + c$$
$$\leq b(e - dr) + c .$$

Thus

$$\dim H^o(\hat{X},\mathcal{L}) \leq \sum_{r=0}^{\frac{e-a}{d}} (b(e-dr) + c) \binom{r+n-2}{n-2} .$$

But $b(e - dr) + c$ is linear in e and r, and $\binom{r+n-2}{n-2}$ is a polynomial of degree n-2 in r. So the expression on the right, which is obtained by numerical integration and substituting $\frac{e-a}{d}$

for r, is a numerical polynomial $P(e)$ of degree n in e. This establishes the lemma.

Proof of the theorem. First we will show that $\operatorname{tr.deg.} K(\hat{X})/k = n$. Indeed, let $\xi_1,\ldots,\xi_r \in K(\hat{X})$ be algebraically independent over k. We shall prove that $r \leq n$.

Given a $\xi \in K(\hat{X})$, we claim that there exist an invertible sheaf $\mathcal{L}$ and sections s_o, s_1 $H^o(\hat{X},\mathcal{L})$ such that $\xi = s_1/s_o$. To see this, let $\mathcal{P} = \{g \in \mathcal{K}_{\hat{X}} \mid g\xi \in \mathcal{O}_{\hat{X}}\}$ be the pole sheaf of ξ, and let $\mathcal{Q} = \xi\mathcal{P}$. We have $\xi\colon \mathcal{P} \longrightarrow \mathcal{Q}$ is an isomorphism. Take $\mathcal{L} \cong \mathcal{P}^{-1} \cong \mathcal{Q}^{-1}$. Now $1 \in \mathcal{P}^{-1}$ and $1 \in \mathcal{Q}^{-1}$ give two sections s_o, s_1 of $\mathcal{L}$ with $\xi = s_1/s_o$.

Now, for each $i = 1,\ldots,r$, let $\mathcal{L}_i$ be an invertible sheaf with sections $s_{oi}, s_{1i} \in H^o(\hat{X},\mathcal{L}_i)$ such that $\xi_i = s_{1i}/s_{oi}$. Let $\mathcal{L} = \mathcal{L}_1 \otimes \ldots \otimes \mathcal{L}_r$. Then $\mathcal{L}$ has non-zero sections $s_o,\ldots,s_r$ such that $\xi_i = s_i/s_o$, $i = 1,\ldots,r$, obtained by forming suitable tensor products of the sections s_{ji}, $j = 0,1$.

Consider the ring

$$A = \sum_{m\geq 0} H^o(\hat{X},\mathcal{L}^m) .$$

We have a natural homomorphism of graded rings

$$A \longrightarrow K(\hat{X})[T] ,$$

where T is an indeterminate, defined by $s \longmapsto (s/s_o^m)T^m$ for any s $H^o(\hat{X},\mathcal{L}^m)$. Clearly this is injective, and so A is a graded subring of $K(\hat{X})[T]$. Thus $s_o,\ldots,s_r$ are algebraically independent over k because their images $T, \xi_1 T,\ldots,\xi_r T$ in $K(\hat{X})[T]$ are algebraically independent over k. Hence

$$\text{tr. deg. } K(A)/k \geq r + 1$$

where $K(A)$ denotes the quotient field of A.

To complete the proof, we need the following lemma.

<u>Lemma 1.9</u>. Let $A = \sum_{m \geq 0} A_m$ be a graded k-algebra, which is an integral domain, and let $K(A)$ be the quotient field of A. Suppose there is a numerical polynomial $P(z) \in \mathbb{Q}[z]$ of degree n, such that

$$\dim_k A_m \leq P(m)$$

for all $m >> 0$. Then

a) tr. deg. $K(A)/k \leq n+1$, and

b) if tr. deg. $K(A)/k = n+1$, then $K(A)$ is a finitely generated field extension of k.

<u>Proof</u>. The proof is easy and will be left as an exercise for the reader. Or, see Hartshorne [CDAV, Lemma 6.3, p. 435].

<u>Proof of the theorem, continued</u>. Now by Lemma 1.8, we have $\dim_k A_m \leq P(m \deg \mathcal{L})$, where P is a numerical polynomial of degree r. Hence by the above lemma, we conclude that tr. deg. $K(A)/k \leq n+1$. Hence $r \leq n$, and tr. deg. $K(A)/k \leq n$. On the other hand, $K(\hat{X})$ contains $K(X)$ whose transcendence degree is n. So we have tr. deg. $K(\hat{X})/k = n$.

Now suppose $\xi_1, \ldots, \xi_n$ form a transcendence basis of $K(\hat{X})$ over k, and let A be the ring constructed from them as above. Our proof shows also that tr. deg. $K(A)/k = n+1$. But then by Lemma 1.9 (b), $K(A)$ is a finitely generated field extension of k.

We claim that $K(A) = K(\hat{X})(T)$. Indeed, we can write A as

$$A = H^0(\hat{X}, \sum_{m\geq 0} \mathcal{L}^m)$$

and

$$K(\hat{X})[T] = H^0(\hat{X}, \mathcal{K}_{\hat{X}}[T]) .$$

Since X is non-singular, $\mathcal{O}_{\hat{X}}$ is integrally closed in $\mathcal{K}_{\hat{X}}$. It follows that $\mathcal{O}_{\hat{X}}[T]$ is integrally closed in $\mathcal{K}_{\hat{X}}[T]$. But locally, $\sum_{m\geq 0} \mathcal{L}^m$ is isomorphic to the sheaf $\mathcal{O}_{\hat{X}}[T]$. Hence A is integrally closed in $K(\hat{X})[T]$. Now since both have the same transcendence degree, their quotient fields must be equal. Hence $K(\hat{X})$ (T) and its subfield $K(\hat{X})$ are finitely generated extensions of k. Therefore, it follows that $K(\hat{X})$ is a finite algebraic extension of $K(X)$, which means that Y is G2 in X.

Remark. The same argument, suitably refined, can be used to prove the following theorem of Hartshorne [CDAV], Theorem 6.7 and Corollary 6.8, pp. 438-440. We will not use this result.

Theorem. Let X be a complete non-singular variety, and let $Y \subseteq X$ be a local complete intersection whose normal bundle $N_{Y/X}$ is ample. Then Y is G2 in X.

The following example shows that the above result cannot be strengthened to G3.

Example. Let $f: X' \longrightarrow X$ be a finite étale morphism of the projective varieties X' and X. Let Y' be a sufficiently general complete intersection in X' with $1 \leq \dim Y' < \frac{1}{2} \dim X'$. Let $Y = f(Y')$. Then Y is G2 but not G3 in X.

For, by Corollary 1.4, Y' is G3 in X'. (If $\dim Y' = 1$, use Corollary 2.3 below instead.) On the other hand, by Exercise 4.10, below, $f|_{Y'}$ is an isomorphism onto Y. Hence $\hat{f}: \hat{X}' \longrightarrow \hat{X}$ is an isomorphism since f is étale. Thus $K(\hat{X}') = K(\hat{X})$. But $K(X) \neq K(\hat{X})$. This shows that Y is G2 but not G3 in X.

Note also that $N_{Y/X} \cong N_{Y'/X'}$ via the isomorphism $f|_{Y'}$. Thus $N_{Y/X}$ is ample because Y' is a complete intersection in X'.

§2. A result of Speiser.

In this section, we give some relations between the properties G2, G3 and Lef of a subvariety, and the cohomological dimension of the complement of that subvariety.

Proposition 2.1. Let X be a complete non-singular variety, and let $Y \subseteq X$ be a closed subscheme. Then Y is G3 in $X \Longleftrightarrow Y$ is G2 in X and Lef (X,Y).

Proof. Suppose Y is G3 in X. Then obviously Y is G2 in X. To see Lef (X,Y), suppose U is an open neighborhood of Y and L an invertible sheaf on U. Choose an embedding of L in the function field $K(X)$, considered as a constant sheaf. We have a commutative diagram

Since X is non-singular, for any $x \in X$, we have $\mathcal{O}_{\hat{X},x} \cap K(X) = \mathcal{O}_{X,x}$ (Exercise 4.12, below). Hence every section $s \in H^0(\hat{X},\hat{L})$ extends to a section of L in a suitable neighborhood U' of Y with $Y \subseteq U' \subseteq U$.

Now for an arbitrary locally free sheaf E on U, take a resolution

$$\oplus M_j \longrightarrow \oplus L_i \longrightarrow E^{\vee} \longrightarrow 0$$

where the L_i and M_j are invertible sheaves on U. (This is possible since U is non-singular). Then we have an exact sequence

$$0 \longrightarrow E \longrightarrow \oplus L_i^{\vee} \longrightarrow \oplus M_j^{\vee} .$$

Now using the result we have just proved for invertible sheaves, and the 5 lemma, one sees easily that there is a neighborhood U' of Y with $U' \subseteq U$, and an isomorphism

$$H^o(U',E) \xrightarrow{\cong} H^o(\hat{X},\hat{E}) .$$

This gives Lef (X,Y).

Conversely, suppose Y is G2 in X and Lef (X,Y). Take an invertible sheaf L on X and sections $s_o,\ldots,s_r \in H^o(X,L)$ such that $\xi_i = s_i/s_o \in K(X)$ generate the field $K(X)$ over k. Consider the ring

$$A = \sum_{m\geq 0} H^o(\hat{X}, \hat{L}^m).$$

By Lef (X,Y), for each m, there exists a neighborhood $U_m \supseteq Y$ such that

$$H^o(U_m,L^m) \xrightarrow{\cong} H^o(\hat{X}, \hat{L}^m)$$

is an isomorphism. Therefore $A \subseteq K(X)[T]$ via the map $s \longmapsto (s/s_o^m)T^m$ for $s \in H^o(\hat{X},\hat{L}^m)$. On the other hand, since the ξ_i generate $K(X)$, we have $K(A) \cong K(X)(T)$. Now as in the proof of Theorem 1.6, we see that A is integrally closed in $K(\hat{X})[T]$. Since Y is G2 in X, tr. deg. $K(X)/k$ = tr. deg. $K(\hat{X})/k$, and so we have $K(A) \cong K(\hat{X})(T)$. Therefore $K(X) = K(\hat{X})$ and Y is G3 in X.

Corollary 2.2. (Speiser [1]). Let X be a complete non-singular variety, and let Y be a closed subscheme. Then the following conditions are equivalent.

i) Y is G3 in X and Y meets every effective divisor on X,

ii) Y is G2 in X and $cd(X-Y) < n-1$.

Proof. Combine the previous Proposition with Proposition 1.1, Chapter IV.

Corollary 2.3. Let X be a non-singular projective variety of dimension n, and let Y be an irreducible curve which is a complete intersection in X. Then Y is G3 in X. (We have already seen, in Corollary 1.4 above, that complete intersections of dimension ≥ 2 are G3.)

Proof. By Corollary 1.2, Chapter IV, we have Lef (X,Y). Since Y is a complete intersection, by Example 4, §4, Chapter III, the normal bundle $N_{Y/X}$ is ample. In fact, we can say more. Let $Y = X \cap H_1 \cap \ldots \cap H_{n-1}$, where the H_i are hypersurfaces in the ambient projective space. Replacing each H_i by a suitable multiple, if necessary, we may assume they all have the same degree, and so correspond to the same ample invertible sheaf L on the projective space. Then $N_{Y/X} \cong (L|_Y)^{\oplus(n-1)}$. Now by Theorem 1.6, Y is G2 in X. Hence by Proposition 2.1, Y is G3 in X.

§3. Theorem of Hironaka and Matsumura.

In this section we give a theorem of Hironaka and Matsumura which characterizes those subvarieties of the projective space $\mathbb{P}^n_k$ that are G3. Also we give a theorem of Hartshorne characterizing those subvarieties of $\mathbb{P}^n_k$ whose complements have cohomological dimension $< n-1$. Finally, we give some new examples of non-algebraizable formal schemes.

Theorem 3.1. (Hironaka-Matsumura [1], Theorem 3.3, p. 69). Let $Y \subseteq \mathbb{P} = \mathbb{P}^n_k$ be a closed subscheme. Then Y is G3 in $\mathbb{P} \Longleftrightarrow Y$ is connected and of dimension ≥ 1.

Proof. Suppose Y is G3 in $\mathbb{P}$. Then by Remark 2 before Proposition 1.1, Y is connected. It also follows that Y has dimension ≥ 1.

Conversely, suppose Y is connected and of dimension ≥ 1. Then we will show that Y is G3 in $\mathbb{P}$.

First, let Y_o be an irreducible curve in Y. Then we have morphisms of formal schemes $\mathbb{P}_{/Y_o} \longrightarrow \mathbb{P}_{/Y} \longrightarrow \mathbb{P}$ (where $X_{/Y}$ denotes the formal completion of X along the closed subscheme Y), and hence we have inclusion of fields

$$K(\mathbb{P}) \subseteq K(\mathbb{P}_{/Y}) \subseteq K(\mathbb{P}_{/Y_o}) .$$

Thus it will be sufficient to show that $K(\mathbb{P}_{/Y_o}) \cong K(\mathbb{P})$. In other words, we may assume that Y is an irreducible curve.

Choose a linear subspace $L^{n-2} \subseteq \mathbb{P}$ of dimension n-2 which does not meet Y. Let $V = \mathbb{P} - L$, and let $\pi\colon V \longrightarrow \mathbb{P}^1$ be the projection with center L onto a suitable line $\mathbb{P}^1 \subseteq \mathbb{P}$. Let Y' be the normalization of Y, and let $f: Y' \longrightarrow \mathbb{P}^1$ be the composition of π with the natural map of Y' to Y.

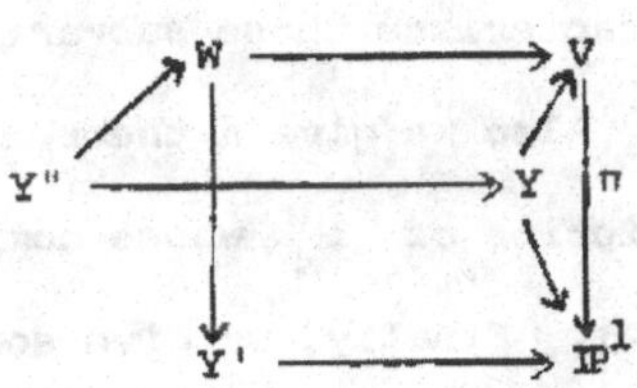

Let $W = V \times_{\mathbb{P}^1} Y'$. Let Y'' be the "diagonal" section of $Y' \longrightarrow W$, constructed as the product of $id: Y' \longrightarrow Y'$ and the natural maps $Y' \longrightarrow Y \longrightarrow V$.

Now V is a geometric vector bundle over $\mathbb{P}^1$. In the notation of Grothendieck [EGA, II, 1.7.8], $V \cong \mathbb{V}\,(\mathcal{O}_{\mathbb{P}^1}(-1)^{\oplus n-1})$. Hence W is a geometric vector bundle over Y'. There is an automorphism of the scheme W, considered as an affine bundle over Y', sending Y'' to the zero-section of W. The latter is a complete intersection. Hence by Corollary 2.3 above, Y'' is G3 in W.

There is a natural map of formal schemes $W_{/Y''} \longrightarrow V_{/Y} = \mathbb{P}_{/Y}$, hence an inclusion of fields

$$K(\mathbb{P}_{/Y}) \subseteq K(W_{/Y''}) = K(W).$$

Thus the branch locus B of the field extension $K(\mathbb{P}_{/Y})/K(\mathbb{P})$ is contained in the branch locus of $K(W)/K(\mathbb{P})$, which is just π^{-1} (branch locus of Y' over $\mathbb{P}^1$).

This argument is independent of the choice of the projection center L^{n-2} and the line $\mathbb{P}^1$. As these vary, the branch locus B of $K(\mathbb{P}_{/Y})/K(\mathbb{P})$ is always contained in the subset π^{-1} (branch locus of Y' over $\mathbb{P}^1$). Thus B must have codimension ≥ 2, so by the purity of branch locus, it is empty. But $\mathbb{P}^n$ is simply connected.

(Exercise 4.11, below). So $K(\mathbb{P}_{/Y}) = K(\mathbb{P})$ which was to be proved.

Theorem 3.2. (Hartshorne [CDAV], Theorem 7.5, p. 444). Let $Y \subseteq \mathbb{P} = \mathbb{P}^n_k$ be a closed subscheme. Then the following are equivalent.

1) Y is connected and of dimension ≥ 1,

2) $\mathrm{cd}\,(\mathbb{P}-Y) < n-1$.

Proof. $(1) \Longrightarrow (2)$:

Since $\dim Y \geq 1$, Y meets every hyperplane in $\mathbb{P}$. Now by Corollary 2.2 and the fact that Y is $G3$ in $\mathbb{P}$, we have $\mathrm{cd}\,(\mathbb{P}-Y) < n-1$.

$(2) \Longrightarrow (1)$:

By Corollary 3.9, Chapter III, Y is connected. Then by the example before Corollary 3.9, Chapter III, Y has dimension ≥ 1.

Example 3.3. Some non-algebraizable formal schemes.

Given a formal scheme $\mathfrak{X}$, we say that $\mathfrak{X}$ is algebraizable if there exists a scheme X, and a closed subscheme $Y \subseteq X$ such that $\mathfrak{X} \cong \hat{X}$, the formal completion of X along Y.

Here we give some new examples of non-algebraizable formal schemes. For other examples, see Hironaka and Matsumura [1, §5].

Let Y be a non-singular subvariety of $\mathbb{P} = \mathbb{P}^n_k$ such that Y is not simply connected (for example, any non-singular curve of genus > 0). Let $f\colon Y' \longrightarrow Y$ be a non-trivial finite étale covering of Y. Then by a result of Grothendieck ([SGA 1], exposé I,

Cor. 8.4), there exists a formal scheme $\mathfrak{X}'$ having Y' as its reduced scheme of definition, and a finite étale morphism $g\colon \mathfrak{X}' \longrightarrow \hat{\mathbb{P}}$, where $\hat{\mathbb{P}} = \mathbb{P}_{/Y}$ is the formal completion of $\mathbb{P}$ along Y, such that the diagram

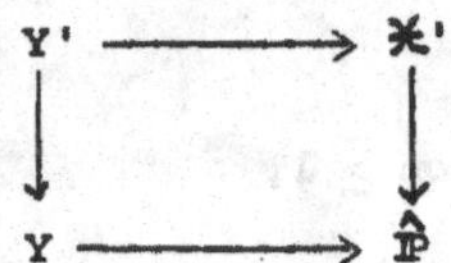

is commutative.

Now we will use the projection method of Hironaka to show that $K(\mathfrak{X}') = K(\mathbb{P})$. Thus $\mathfrak{X}'$ cannot be algebraizable.

Let $\dim Y = r$. Choose a linear subspace L^{n-r-1} which does not meet Y, and project from L^{n-r-1} onto a suitable $\mathbb{P}^r \subseteq \mathbb{P}$. Let $V = \mathbb{P} - L^{n-r-1}$. Then the projection $\pi\colon V \longrightarrow \mathbb{P}^r$ makes V into a vector bundle $V = \mathbb{V}(\mathcal{O}_{\mathbb{P}}(-1)^{\oplus n-r})$, and $\pi|_Y$ is a finite morphism from Y to $\mathbb{P}^r$. Make the base extension $f\colon Y' \longrightarrow \mathbb{P}^r$, and let Y'' be the diagonal section of the bundle $W = f^{-1}(V)$. Then the embedding of Y'' in W is isomorphic to the embedding of Y' as zero section; this is a complete intersection, so Y'' is G3 in W. Now the natural map $W_{/Y''} \longrightarrow V_{/Y}$ factors through $\mathfrak{X}'$, since $\mathfrak{X}'$ is étale over $V_{/Y}$ [SGA 1], exposé I, Thm. 5.5. Thus $K(\mathfrak{X}') \subseteq K(W_{/Y''}) = K(W)$, and the branch locus of $K(\mathfrak{X}')$ over $\mathbb{P}$ must be contained in π^{-1} (branch locus of $Y'/\mathbb{P}^r$). Now by shifting the positions of L^{n-r-1} and $\mathbb{P}^r$, we find, by the theorem of purity of the branch locus, that $K(\mathfrak{X}')$ is unramified over $K(\mathbb{P})$. Whence $K(\mathbb{P}) = K(\mathfrak{X}')$ since $\mathbb{P}$ is simply connected (Exercise 4.11, below).

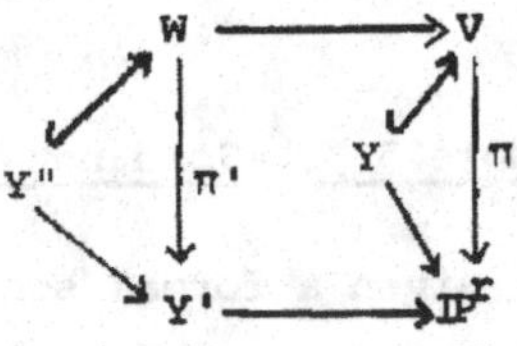

$$\begin{array}{ccccc} & & W & \longrightarrow & V \\ & \nearrow & \downarrow{\scriptstyle \pi'} & Y \nearrow & \downarrow{\scriptstyle \pi} \\ Y'' & & & & \\ & \searrow & Y' & \longrightarrow & \mathbb{P}^r \end{array}$$

§4. Exercises and Problems.

Exercise 4.1. Let X be a complete non-singular variety, and let $Y \subseteq X$ be a connected closed subscheme of dimension ≥ 1. Assume that Y is locally a complete intersection and that $N_{Y/X}$ is ample. Then prove that

a) the ring $H^{o}(\hat{X}, \mathcal{O}_{\hat{X}})$ of formal-regular functions along Y is isomorphic to k, and

b) for any morphism $f: X \longrightarrow X'$, $\dim f(X) \geq 1 \Longrightarrow \dim f(Y) \geq 1$.

Exercise 4.2. Give an example of a non-singular projective variety X containing a non-singular subvariety of codimension 1 such that Y is G2 but not G3 in X.

Exercise 4.3. Let X be a complete non-singular variety, and let $Y \subseteq X$ be a connected closed subscheme. Let $Y_1, \ldots, Y_r$ be the irreducible components of Y. Assume for some i, Y_i is G2 (resp. G3) in X. Then prove that Y is G2 (resp. G3) in X.

Exercise 4.4. Give an example to show that the converse of 4.3 is false. In fact, find a complete non-singular surface X containing two irreducible curves Y_1, Y_2 such that Y_1 and Y_2 are not G2 in X, but $Y = Y_1 \cup Y_2$ is G3 in X.

Exercise 4.5. Let X be a complete non-singular surface, and let $Y \subseteq X$ be a reduced curve having its irreducible components $Y_1, \ldots, Y_r$. Let $a_{ij} = (Y_i . Y_j)$. Give a necessary and sufficient condition on the intersection matrix (a_{ij}) so that

a) $q(X-Y) = 0$

b) $p(X-Y) = 1$

c) Lef (X,Y)

d) Y is G2 in X

e) Y is G3 in X

f) there exists an effective Cartier divisor D with Supp $D = Y$, and $\mathcal{O}_D(D)$ is ample on D.

g) there exists an effective Cartier divisor D with Supp $D = Y$, and $\mathcal{O}_D(-D)$ is ample on D.

In particular determine which of these conditions are equivalent to each other. If Y is irreducible, show that

a), c), d), e), f) $\Longleftrightarrow (Y^2) > 0$, and b), g) $\Longleftrightarrow (Y^2) < 0$.

<u>Exercise 4.6</u>. Let X be a complete non-singular variety and let $Y \subseteq X$ be a closed subscheme. Then show that $q(X-Y) = 0 \Longrightarrow Y$ is G3 in X.

<u>Problem 4.7</u>. Let X be a complete-singular variety of dimension n, and let $Y \subseteq X$ be a closed subscheme. Then is it true that "Y is G2 in $X \Longleftrightarrow q(X-Y) < n-1$"?

<u>Problem 4.8</u>. Let Y be a closed subscheme of a projective variety X of dimension n such that $\dim Y \geq \frac{n}{2}$ and $N_{Y/X}$ is ample. Then is Y G3 in X? (Assume char. $k = 0$ if necessary.)

Exercise 4.9. Let Y be a connected closed subscheme of a normal variety X, and assume that Y is G2 in X. Let X' be the normalization of X in $K(\hat{X})$, and let $f: X' \longrightarrow X$ be the natural map. Show that there is a subvariety Y' of X' which is G3 in X', such that $f|_{Y'}$ is an isomorphism of Y' onto Y, and such that f is étale at points of X' in a suitable neighborhood of Y'. (We say that (X',Y') is an étale neighborhood of (X,Y).)

Exercise 4.10. Let $f: X' \longrightarrow X$ be a finite étale morphism of the projective varieties X' and X. Let Y' be a sufficiently general complete intersection in X' with $\dim Y' < \frac{1}{2} \dim X'$. Let $Y = f(Y')$. Then prove that $f|_{Y'}$ is an isomorphism onto Y.

Exercise 4.11. Prove that the projective space $\mathbb{P}^n_k$ is simply connected, i.e., has no finite étale covers.

Exercise 4.12. Let A be a regular local ring. Let $K(A)$ be the quotient field of A, and let $\hat{A}$ be the completion of A. Then prove that

$$K(A) \cap \hat{A} = A .$$

(In fact, this is true for any local ring A such that A and $\hat{A}$ are UFD).

CHAPTER VI

ALGEBRAIC GEOMETRY AND ANALYTIC GEOMETRY

§1. Analogous results in analytic geometry.

Many of the questions we have been studying in this seminar have analogues in the theory of several complex variables. Algebraic geometry and "analytic geometry" have had a parallel development. The two subjects mutually enrich each other by suggesting new results. In this section we will present a brief survey, without proofs, of analytic results analogous to the algebraic results in the earlier parts of the course.

(i) Positive line bundles.

Kodaira [1] give a differential-geometric definition of positive line bundle on a complex Kähler variety. For algebraic varieties, one can show that this coincides with our notion of ampleness. He then proves the following

Theorem (Kodaira's Vanishing Theorem). Let L be a positive line bundle on a compact Kähler manifold X. Then

$$H^i(X,L^\vee) = 0 \qquad \text{for} \quad i < \dim X$$

and

$$H^i(X,L \otimes \Omega^n_X) = 0 \qquad \text{for} \quad i > 0 .$$

At the moment, no algebraic proof of the analogous statement for algebraic varieties is known. However, Mumford [2] has shown that the theorem fails for algebraic varieties over a field of characteristic p (see his Example 6, p. 102). He also gives an algebraic proof of the vanishing of $H^1(X,L^V)$ for dim $X \geq 2$ (his Theorem 2, p. 95).

(ii) Stein spaces.

Stein spaces are the analogues of affine algebraic varieties.

We recall the definition of a Stein space (Gunning and Rossi [1], pp. 208, 209). An analytic space X is said to be a Stein space if it has a countable topology, is holomorphically convex, and its global holomorphic functions separate points, and generate the maximal ideal $\mathfrak{m}_x$ for every point $x \in X$.

Stein spaces can be characterized by cohomological properties, or by the existence of a global function with suitable convexity properties.

Theorem. Let X be an analytic space. Then the following conditions are equivalent.

(i) X is Stein,

(ii) $H^i(X,F) = 0$ for all $i > 0$ and all coherent analytic sheaves F,

(iii) X is 1-complete in the sense of Andreotti and Grauert (see definition below).

Proof. The equivalence of (i) and (ii) is a theorem of Serre (see Gunning and Rossi [1], §A, Theorem 14, p. 243 and Theorem 20, p. 246, Chapter VIII). The equivalence of (i) and (iii) is the so-called "Levi problem", solved by R. Narasimhan [1], p. 355.

There is an analogous result for modifications of Stein spaces.

Theorem. Let X be an analytic space. Then the following conditions are equivalent.

(i) X is holomorphically convex, and is a modification of a Stein space at finitely many points,

(ii) $H^i(X,F)$ is finite-dimensional for all $i > 0$ and all coherent sheaves F on X,

(iii) X is 1-pseudoconvex in the sense of Andreotti and Grauert (see definition below).

Proof. The equivalence of (i) and (ii) is proved by R. Narasimhan [2], Theorem V, p. 196. (i)$\Longrightarrow$(iii) follows easily from the implication (i)$\Longrightarrow$(iii) of the previous theorem, and (iii)$\Longrightarrow$(ii) follows from the finiteness theorem of Andreotti and Grauert [1], quoted below.

(iii) Vector bundles.

For line bundles it seems that all possible notions of positivity coincide. However, for vector bundles of rank > 1, there are many notions analogous to our notion of ample vector bundle. The analytic notion closest to ours is what Grauert [2] calls weakly positive (his definition p. 342). Some other notions have been introduced by Griffiths [1], which he calls weakly positive (different from Grauert's), positive, and ample (different from ours). Griffiths [2] studies subvarieties with weakly positive normal bundle, and proves a theorem of extension of cohomology classes for them.

(iv) Finiteness and vanishing theorems for cohomology.

Andreotti and Grauert [1] have proved some basic finiteness and vanishing theorems for cohomology on non-compact analytic spaces (see also Grauert [1]). We recall their definitions (pp. 216, 235, 236) of q-pseudoconvexity, q-completeness and q-pseudoconcavity.

Let $q \geq 1$ be an integer. A real-valued C^{∞}-function f on a domain $D \subseteq \mathbb{C}^n$ is strictly q-pseudoconvex if its Levi form

$$L(f) = \sum \frac{\partial^2 f}{\partial z_i \partial \bar{z}_j} u_i \bar{u}_j$$

has at least $n-q+1$ positive eigenvalues. A real-valued C^{∞}-function f on an analytic space X is strictly q-pseudoconvex, if it is locally the restriction to X of a strictly q-pseudoconvex function on an open domain $D \subseteq \mathbb{C}^n$ for some n. (See also Andreotti [1], p. 13, for the independence on the local embedding).

A complex space X is strictly q-pseudoconvex if there is a compact set $K \subseteq X$, and a continuous real-valued function f on X, which is strictly q-pseudoconvex outside K, and such that for each $c \in \mathbb{R}$, the set

$$B_c = \{x \in X \mid f(x) < c\}$$

is relatively compact in X.

A complex space X is q-complete if it is possible to take K empty in the above definition.

A complex space X is strictly q-pseudoconcave if there is a compact set $K \subseteq X$ and a real-valued continuous function $f > 0$ on X, which is strictly q-pseudoconvex outside K, and such that for all $c > 0$, $c \in \mathbb{R}$, the set

$$B_c = \{x \in X \mid f(x) > c\}$$

is relatively compact.

Theorem (Andreotti and Grauert [1], Theorem 14, p. 249 and Corollary, p. 250). Let X be an analytic space of dimension n. Then

a) if X is strictly q-pseudoconvex, $H^i(X,F)$ is finite-dimensional for all $i \geq q$, and all coherent sheaves F.

b) if X is q-complete, then $H^i(X,F) = 0$ for all $i \geq q$, and all coherent sheaves F.

c) if X is strictly q-pseudoconcave, then $H^i(X,F)$ is finite-dimensional for all $i < n-q$, and all locally free sheaves F.

Problem 1.1. By analogy with algebraic theorems, it seems likely that suitable conditions on a subspace of a compact analytic space should imply convexity or concavity of the complement. Specifically, let X be a compact complex manifold of dimension n, and let Y be an analytic subspace of dimension s. Let $U = X - Y$.

a) Show that $X - Y$ is strictly (s+1)-pseudoconcave.

b) Assume that Y is a submanifold whose normal bundle is ample (in some sense). Show that $X - Y$ is strictly (n-s)-pseudoconvex.

The cohomological conditions which arise in the theorem of Andreotti and Grauert above have also been studied by Sorani and Villani [1], and Villani [1]. Vesentini [1], Theorem 4.2, p. 106, gives another proof of part b of the above theorem.

(v) Subvarieties of projective space.

For subspaces of projective space, fairly complete results on cohomology have been obtained by Barth [2],[3] and [4]. In his paper [2], he defines a metric on the projective space $\mathbb{P} = \mathbb{P}^n_{\mathbb{C}}$, and uses the distance from a submanifold to verify the conditions of pseudo-convexity of Andreotti and Grauert.

Theorem. (Barth [2], Satz 3). Let Y be a closed s-dimensional submanifold of $\mathbb{P} = \mathbb{P}^n_{\mathbb{C}}$. Then $U = \mathbb{P} - Y$ is strictly (n-s)-pseudoconvex and strictly (s+1)-pseudoconcave.

Corollary. With the same hypotheses, $H^i(U,F)$ is finite-dimensional

a) for $i \geq n-s$, and all coherent analytic sheaves F on U, and

b) for $i < n-s-1$, and all locally free analytic sheaves F on U.

Theorem (Barth [3]). Let Y be a closed connected submanifold of dimension ≥ 1 of $\mathbb{P} = \mathbb{P}^n_{\mathbb{C}}$, and let $U = \mathbb{P} - Y$. Then $H^{n-1}(U,F) = 0$ for all coherent analytic sheaves on U.

Finally, he gives the following criterion for the vanishing of the analytic cohomology groups of coherent algebraic sheaves (by Serre's theorems [GAGA], the condition that F be coherent analytic on all of $\mathbb{P}$ implies that it comes from a coherent algebraic sheaf - see §2 below).

Theorem (Barth [4], §6, Theorem II). Let Y be a closed submanifold of dimension s of $\mathbb{P} = \mathbb{P}^n_{\mathbb{C}}$, let $U = \mathbb{P} - Y$, and let $r \geq n-s$ be an integer. Then the following conditions are equivalent.

(1) $H^i(U,F|_U) = 0$ for all $i \geq r$, and all coherent analytic sheaves F on $\mathbb{P}$,

(2) $H^i(U,\omega) = 0$ for all $i \geq r$, where $\omega = \Omega^n_U$,

(3) The natural maps

$$H^i(\mathbb{P},\mathbb{C}) \longrightarrow H^i(Y,\mathbb{C})$$

are isomorphisms for all $i < n-r$.

Theorem (Barth [4], §7, Corollary to Theorem III). With the hypotheses of the previous theorem, the conditions (1), (2), (3) always hold for $r = 2n - 2s - 1$.

Remarks. 1. This last theorem gives non-trivial results in case $\dim Y \geq \frac{1}{2}n$. In that case it generalizes a well-known theorem of Lefschetz (see Chapter IV, §4 above) to subspaces which need not be complete intersections. It can also be interpreted as giving obstructions to embedding a manifold Y in projective space. For example, if the first Betti number of Y is non-zero, then Y cannot be embedded in $\mathbb{P}^{2s-1}_{\mathbb{C}}$.

2. We still do not know whether the cohomology groups $H^i(U,F)$ vanish under the above conditions for every coherent analytic sheaf on U.

(vi) Extension of coherent analytic sheaves.

In algebraic geometry, a coherent sheaf defined on an open set always extends to a coherent sheaf on the whole space [EGA, I, §9]. The analogous statement for coherent analytic sheaves is false. Serre [6] gave examples of analytic line bundles on $\mathbb{C}^2 - \{(0,0)\}$ which do not extend to the whole plane (see Exercise 2.8 below). In the same paper he posed the problem of extending a coherent analytic sheaf from an open set to the whole space. Also he asked for conditions under which the analytic local cohomology sheaves $\underline{H}_Y^i(F)$ would be coherent, and he asked when the algebraic local cohomology sheaves would generate the analytic ones. These three questions have recently been answered independently by Frisch and Guenot, Siu, and Trautmann. We will give their main results in a slightly simplified form. For more precise necessary and sufficient conditions, see the articles referred to.

Theorem (Extension Theorem. Frisch-Guenot [1], Corollary VII.5, p. 342 and Siu [1], Theorem 2, p. 108). Let X be an analytic space, let $Y \subseteq X$ be a closed analytic subset, and let F be a coherent analytic sheaf on $U = X - Y$. Assume depth $F \geq \dim Y + 3$. Then i_*F is coherent on X, where $i: U \longrightarrow X$ is the inclusion.

Theorem (Coherence Theorem. Siu [2], Theorem A, and Trautmann [2], §III, Theorem 2.1, p. 157). Let X be an analytic space, let Y be an analytic subspace, and let F be a coherent analytic sheaf on X. Then the local cohomology sheaves $\underline{H}_Y^i(F)$ are coherent for all $i <$ depth $F - \dim Y$.

<u>Theorem</u> (<u>Comparison Theorem</u>. Siu [2], Theorem B). Let X be an algebraic variety over $\mathbb{C}$, let Y be a closed subvariety, and let F be a coherent algebraic sheaf on X. Let "h" denote the functor "associated analytic space" (see §2, below). Let q be an integer. Then the $\underline{H}^i_Y(F)$ are coherent (algebraic) sheaves for all $i \leq q$ if and only if the $\underline{H}^i_{Y^h}(F^h)$ are coherent (analytic) sheaves for all $i \leq q$. Furthermore, in that case, the natural maps

$$\underline{H}^i_Y(F)^h \longrightarrow \underline{H}^i_{Y^h}(F^h)$$

are isomorphisms for all $i \leq q$.

<u>Remark</u>. One obtains a special case of the above theorems by assuming $X-Y$ to be a manifold, and $F|_{X-Y}$ locally free. In that case depth $F = n = \dim X$.

(vii) <u>Meromorphic functions</u>.

If Y is an analytic subset of an irreducible analytic space X, we can study the field of meromorphic functions defined in a neighborhood of Y. More generally, we can study the field of meromorphic functions on a non-compact space.

Andreotti [1] defines the notion of a <u>pseudoconcave</u> analytic space (p. 21). In particular, it follows from his Proposition 10, p. 22, that if an n-dimensional analytic space is strictly q-pseudoconcave for any $q \leq n-1$, then it is pseudoconcave. He proves

<u>Theorem</u> (Andreotti [1], Theorem 5, p. 35). Let X be a normal irreducible pseudoconcave analytic space. Then the field of meromorphic functions $K(X)$ is a finitely generated extension field of $\mathbb{C}$, of transcendence degree $\leq \dim X$.

Chow [3] proves some theorems for homogeneous spaces, which are analogous to the theorems of Hironaka and Matsumura. Let X be a <u>homogeneous algebraic variety</u> over $\mathbb{C}$, i.e., a complete non-singular algebraic variety, together with an algebraic group G which acts transitively on X. We say that a subset Y <u>generates</u> X, if the following condition holds: fix a point $y \in Y$. Let G_Y be the subgroup of G generated by $\{g \in G \mid g(y) \in Y\}$. Then G_Y should generate G. (Note that G_Y is independent of the point $y \in Y$).

<u>Theorem</u> (<u>Chow</u> [3], Theorem 2, p. 397). Let X be a homogeneous algebraic variety over $\mathbb{C}$. Let Y be an algebraic subvariety which generates X. Let N be a connected open neighborhood of Y (in the usual topology). Then the field $K(N)$ of meromorphic functions on N is an algebraic extension of the function field $K(X)$.

<u>Theorem</u> (<u>Chow</u> [3], Theorem 3, p. 400). With the same hypotheses and notations, assume furthermore that either

a) X is a rational variety, or

b) $\dim Y \geq \frac{1}{2} \dim X$ and $(Y.Y) \neq 0$ (for algebraic equivalence of cycles).

Then $K(N) = K(X)$.

For example, the second theorem applies to any connected positive-dimensional subvariety of projective space.

Closely related to these results is a theorem of Rossi [1] on extending analytic varieties, defined in a neighborhood of an algebraic subvariety of projective space, to the whole projective space.

Barth [1] has studied meromorphic functions defined in a neighborhood of a subvariety of a complex torus.

§2. Comparison theorems.

Let X be an algebraic variety over $\mathbb{C}$. Then there is a natural way to define an analytic space X^h associated to X (see Serre [GAGA], §2). Roughly speaking, the construction is as follows. Cover X with open affine sets U_i. Embed U_i as a closed subscheme of a suitable affine space $\mathbb{A}^n_{\mathbb{C}}$, and let its ideal be generated by polynomials $f_1,\ldots,f_r$. These polynomials define a closed analytic subspace of $\mathbb{C}^n$, which we call U_i^h. We then glue together the analytic space U_i^h to obtain X^h. Furthermore, to every coherent algebraic sheaf F on X, one can associate functorially a coherent analytic sheaf F^h on X^h, and there are natural maps of cohomology

$$\alpha_i: \ H^i(X,F) \longrightarrow H^i(X^h,F^h)$$

(see Serre [GAGA], §3).

This construction suggests some natural questions. Does every analytic space come from an algebraic variety, and if so, is it unique? (Of course in this generality the answer is no). If X is an algebraic variety, does every coherent analytic sheaf on X^h come from an algebraic sheaf on X, and is it unique? If F is a coherent algebraic sheaf on X, are the maps α_i on cohomology isomorphisms? Again the answers are no in general, as one can see by simple examples (see the exercises at the end of this section).

On the other hand, we have a fundamental theorem of Serre, as generalized by Grothendieck, which says that in the case of complete varieties the answers are Yes.

Theorem (Serre [GAGA], Theorems 1,2,3, pp. 19,20 and Grothendieck [2], Corollary 1, p. 2-09 and Theorem 6, p. 2-14). Let X be a complete algebraic variety over $\mathbb{C}$. Then the functor $F \longmapsto F^h$ induces an equivalence of categories between coherent algebraic sheaves on X and coherent analytic sheaves on X^h. Furthermore, the natural map of cohomology

$$\alpha_i : H^i(X,F) \longrightarrow H^i(X^h,F^h)$$

is an isomorphism for all i, and all coherent sheaves F on X.

Theorem (Chow [1] and Serre [GAGA], Proposition 13, p. 29). Every closed analytic subspace of $\mathbb{P}^n_{\mathbb{C}}$ is algebraic.

Our purpose in this section is to extend the theorem of Serre to certain non-complete algebraic varieties. It turns out that in two cases where we know that the cohomology groups are finite-dimensional, we can prove that the maps α_i are isomorphisms. One theorem is for small values of i, the other for large values of i.

Theorem 2.1. Let X be a complete algebraic variety of dimension n over $\mathbb{C}$, let Y be a subvariety of dimension s, and let $U = X - Y$. Assume that U is non-singular. Then

a) The natural maps

$$\alpha_i : H^i(U,F) \longrightarrow H^i(U^h,F^h)$$

are isomorphisms for all $i < n-s-1$, and all coherent locally free sheaves F on U. Furthermore, these groups are finite-dimensional.

b) If $n-s \geq 2$, then the functor $F \longmapsto F^h$ is fully faithful on the category of locally free sheaves on U.

c) If $n-s \geq 3$, then the functor $F \longmapsto F^h$ induces an equivalence of categories between coherent locally free algebraic sheaves on U and coherent locally free analytic sheaves on U^h.

Proof. By the coherence and comparison theorems of Siu [2] and Trautmann [2] mentioned in §1 (vi) above, we find that the sheaves of local cohomology $\underline{H}^i_Y(F)$ and $\underline{H}^i_Y(F^h)$ are coherent, and the natural maps

$$\underline{H}^i_Y(F)^h \longrightarrow \underline{H}^i_Y(F^h)$$

are isomorphisms, for any coherent algebraic sheaf F on X whose restriction to U is locally free, for all $i < n-s$. We will use the spectral sequence of local cohomology (Grothendieck [LC], Proposition 1.4) in the algebraic and the analytic case. The functor h gives a morphism of spectral sequences

$$\begin{array}{ccc} E_2^{pq} = H^p(X, \underline{H}^q_Y(F)) & \Longrightarrow & H^i_Y(X, F) \\ \downarrow & & \downarrow \\ {}^hE_2^{pq} = H^p(X^h, \underline{H}^q_Y(F^h)) & \Longrightarrow & H^i_Y(X^h, F^h) \end{array} .$$

So by Serre's theorem, the maps $E_2^{pq} \longrightarrow {}^hE_2^{pq}$ are isomorphisms for $q < n-s$, and all values of p. It follows that the map on the abutments is an isomorphism for all $i < n-s$. Furthermore, the groups in this range are all finite-dimensional.

Now we write the long exact sequence of local cohomology in both cases.

$$\begin{array}{ccccccccc} H^i_Y(X,F) & \longrightarrow & H^i(X,F) & \longrightarrow & H^i(U,F) & \longrightarrow & H^{i+1}_Y(X,F) & \longrightarrow & H^i(X,F) \\ \downarrow & & \downarrow & & \downarrow \alpha_i & & \downarrow & & \downarrow \\ H^i_Y(X^h,F^h) & \longrightarrow & H^i(X^h,F^h) & \longrightarrow & H^i(U^h,F^h) & \longrightarrow & H^{i+1}_Y(X^h,F^h) & \longrightarrow & H^i(X^h,F^h) \end{array} .$$

The second and fifth vertical arrows are isomorphisms for all i, by Serre's theorem; the first and fourth are isomorphisms for $i+1 < n-s$. Hence by the five-lemma, the maps α_i are isomorphisms for all $i < n-s-1$. The finite-dimensionality also follows from this sequence. This proves part a) of the theorem.

Part b) follows from part a), for $i = 0$, applied to the sheaf Hom (F,G), for any two locally free sheaves F and G on U. Part c) follows from part b) and the extension theorem of Frisch-Guenot [1] and Siu [1], mentioned in §1 (vi) above.

Remark. One can refine the theorem and its proof using the notion of depth. The theorem remains true, without the hypothesis U non-singular and F locally free, if one replaces the number n by depth F throughout.

Theorem 2.2. Let Y be a non-singular s-dimensional subvariety of $\mathbb{P} = \mathbb{P}^n_{\mathbb{C}}$, and let $U = \mathbb{P} - Y$. Then the maps

$$\alpha_i: \ H^i(U,F) \longrightarrow H^i(U^h,F^h)$$

are isomorphisms for all $i > n-s$, and surjective for $i = n-s$, for all coherent sheaves F on U. Furthermore, these groups are all finite-dimensional.

Before proceeding to the proof, we need to establish a local form of Serre duality, and recall an extension theorem of Griffiths.

<u>Theorem</u> (<u>Serre duality</u>). Let X be a σ-compact complex manifold of dimension n. Let F be a locally free sheaf on X, and let $\omega = \Omega^n_{X/k}$. Then cup-product pairings

$$H^i(X,F) \times H_c^{n-i}(X,F^\vee \otimes \omega) \longrightarrow H_c^n(X,\omega)$$

and the trace map $H_c^n(X,\omega) \longrightarrow \mathbb{C}$ induce a duality homomorphism

$$\varphi_i : H_c^{n-i}(X, F^\vee \otimes \omega) \longrightarrow H^i(X,F)' .$$

If $H^i(X,F)$ is finite-dimensional, then φ_i is surjective and φ_{i-1} is injective.

<u>Proof</u>. This follows immediately from Serre [2], Theorem 2, p. 20 and Proposition 6, p. 21, except that we have refined the statement by analyzing the proof of Lemma 1, p. 19 (loc. cit.). See also Laufer [1] and Suominen [1] for other proofs.

<u>Theorem 2.3</u>. Let X be a compact complex manifold of dimension n. Let Y be a closed subset of X, and let F be a locally free sheaf on X. Then the cup-product and natural maps

$$H_Y^i(X,F) \times H^{n-i}(Y, F^\vee \otimes \omega|_Y) \longrightarrow H_Y^n(X,\omega) \longrightarrow \mathbb{C}$$

induce a duality homomorphism

$$\psi_i : H^{n-i}(Y, F^\vee \otimes \omega|_Y) \longrightarrow H_Y^i(X,F)' .$$

If $H_Y^i(X,F)$ is finite-dimensional, then ψ_i is surjective and

ψ_{i-1} is injective. (Here, for any sheaf G on X, $G|_Y$ denotes the set-theoretic restriction to Y.)

Proof. We consider the exact sequence of cohomology associated to a closed subset (Godement [1], Chapter II, Theorem 4.10.1, p. 190) for the sheaf $G = F^\vee \otimes \omega$, and its natural map by the duality homomorphisms into the dual of the exact sequence of local cohomology (Grothendieck [LC], for the sheaf F (let $U = X - Y$):

$$\begin{array}{ccccccccc}
H_c^{n-i}(U,G) & \longrightarrow & H^{n-i}(X,G) & \longrightarrow & H^{n-i}(Y,G|_Y) & \longrightarrow & H_c^{n-i+1}(U,G) & \longrightarrow & H^{n-i+1}(X,G) \\
\downarrow \varphi_i & & \downarrow \cong & & \downarrow \psi_i & & \downarrow \varphi_{i-1} & & \downarrow \cong \\
H^i(U,F)' & \longrightarrow & H^i(X,F)' & \longrightarrow & H_Y^i(X,F)' & \longrightarrow & H^{i-1}(U,F)' & \longrightarrow & H^{i-1}(X,F)'
\end{array}$$

The second and fifth vertical arrows are isomorphisms by Serre duality for the compact manifold X. Since X is compact, the groups in those columns are finite-dimensional.

Now suppose that $H_Y^i(X,F)$ is finite-dimensional. Then $H^{i-1}(U,F)$ is also finite-dimensional. It follows by Serre duality for U that φ_{i-1} is surjective, and φ_{i-2} is injective. A simple diagram-chase then shows that ψ_i is surjective and ψ_{i-1} is injective.

Remark. This theorem is the analytic analogue of the "formal duality" of Theorem 3.3, Chapter III.

Theorem (Griffiths [2], Theorems I, III, pp. 378-379). Let X be a complex manifold, and let Y be a compact complex submanifold of dimension s. Assume that the normal bundle $N_{Y/X}$ is weakly positive. Then for any locally free sheaf F on X, there is an integer r_o such that for any $r \geq r_o$, the natural map

$$\rho_i: \ H^i(Y,F|_Y) \longrightarrow H^i(Y_r, F \otimes \mathcal{O}_{Y_r})$$

is an isomorphism for all $i < s-1$, and injective for $i = s-1$. (Here $F|_Y$ is the set-theoretic restriction of F to Y; Y_r is the sub-analytic space of X defined by I_Y^r, where I_Y is the sheaf of ideals of Y).

Remark. The definition of weakly positive is given in Griffiths [1], p. 122. In that same paper, he shows that the normal bundle to any submanifold of complex projective space is weakly positive (Corollary, p. 138). This is the case which will interest us.

Proof of Theorem 2.2. Let F be a locally free sheaf on $\mathbb{P} = \mathbb{P}^n_{\mathbb{C}}$, and let $G = F^\vee \otimes \omega$. Let Y_r be the subscheme defined by I_Y^r, where I_Y is the sheaf of ideals of Y. Then we have $(Y_r)^h = (Y^h)_r$. Let $G_r = G \otimes \mathcal{O}_{Y_r}$. Let $\hat{\mathbb{P}}$ be the formal completion of $\mathbb{P}$ along Y. Then we have a commutative diagram

$$\begin{array}{ccccc}
\varprojlim H^{n-i}(Y_r,G_r) & \xleftarrow{(1)} & H^{n-i}(\hat{\mathbb{P}},\hat{G}) & \xrightarrow{(2)} & H^i_Y(\mathbb{P},F)' \\
\downarrow (3) & & & & \uparrow \beta_i' \\
\varprojlim H^{n-i}(Y^h_r,G^h_r) & \xleftarrow{\rho_{n-i}} & H^{n-i}(Y,G^h|_Y) & \xrightarrow{\psi_i} & H^i_Y(\mathbb{P}^h,F^h)'
\end{array}$$

Now (1) is an isomorphism for all i by [EGA, 0_{III}, 13.3.1]. The map (2) is the formal duality isomorphism of Theorem 3.3, Chapter III. The map (3) is the limit of comparison maps, which are isomorphisms by Serre's theorem applied to the projective scheme Y_r. The maps

ψ_i and ρ_{n-i} are those of the local Serre duality and the theorem of Griffiths just stated. Finally, β_i' is the dual of the comparison map of local cohomology

$$\beta_i: H_Y^i(\mathbb{P}, F) \longrightarrow H_Y^i(\mathbb{P}^h, F^h) .$$

We now apply the finiteness theorem of Barth [2] quoted in §1 (v) above: for a locally free sheaf F on $\mathbb{P}$, the cohomology groups $H^i(U^h,F^h)$ are finite-dimensional for all $i \geq n-s$. (By the way, the algebraic cohomology groups $H^i(U,F)$ are also finite-dimensional in this range, by Theorem 5.2, Chapter III). Thus by the long exact sequence of local cohomology, the groups $H_Y^i(\mathbb{P}^h,F^h)$ are finite-dimensional for $i > n-s$. Therefore the duality maps ψ_i above are isomorphisms for $i > n-s$, and injective for $i = n-s$. On the other hand, the homomorphisms ρ_{n-i} of Griffiths' theorem are isomorphisms for $n-i < s-1$, and injective for $n-i = s-1$. We conclude from the diagram above that β_i' is an isomorphism for $i > n-s+1$ and injective for $i = n-s+1$. Hence its dual β_i is an isomorphism for $i > n-s+1$, and surjective for $i = n-s+1$.

Finally, as in the proof of Theorem 2.1 above, we write the long exact sequence of local cohomology in the algebraic and analytic cases, and the comparison homomorphism between them. We conclude by using Serre's theorem on $\mathbb{P}$ that the maps

$$\alpha_i: H^i(U,F) \longrightarrow H^i(U^h,F^h)$$

are isomorphisms for $i > n-s$ and surjective for $i = n-s$, for all locally free sheaves F on $\mathbb{P}$.

Now any coherent algebraic sheaf on U extends to a coherent sheaf on $\mathbb{P}$, which in turn has a resolution by locally free sheaves. Thus by a standard argument using descending induction on i, we find that α_i is an isomorphism (resp. surjective) in the same range for all coherent sheaves on U.

Remark. Except in the finiteness theorem of Barth, we have not used the fact that the ambient space is projective space itself. Thus it seems likely that the same result will hold for X a compact complex manifold of dimension n, and Y a compact submanifold of dimension s with weakly positive normal bundle.

Corollary 2.4. Let Y be a non-singular s-dimensional subvariety of $\mathbb{P} = \mathbb{P}^n_{\mathbb{C}}$, let $U = \mathbb{P} - Y$, and let $r \geq n-s$ be an integer. Then

(i) cd (U) < r implies

(ii) the natural maps

$$H^i(\mathbb{P}^h, \mathbb{C}) \longrightarrow H^i(Y^h, \mathbb{C})$$

are isomorphisms for all $i < n-r$.

Conversely, (ii) implies cd (U) < r+1. Furthermore, (ii) always holds for $r = 2n-2s-1$, and so we always have cd (U) < 2n-2s.

Proof. This follows immediately from Theorem 2.2 above, and the theorem of Barth [4], §6, Theorem II", quoted in §1 (v) above.

In view of the comparison theorems proved in this section, it seems reasonable to make the following conjecture. Presumably such a result will follow from a more natural proof of the comparison theorems.

Conjecture 2.5. Let U be a scheme of finite type over $\mathbb{C}$. For any coherent sheaf F on U, we consider the natural maps

$$\alpha_i: H^i(U,F) \longrightarrow H^i(U^h,F^h).$$

a) If U is non-singular, then α_i is an isomorphism for all $i < p(U)$, and all locally free F.

b) For any U, α_i is an isomorphism for all $i > q(U)$, and all coherent F.

c) If U is non-singular, the functor $F \longmapsto F^h$ on locally free sheaves is fully faithful for $p(U) \geq 1$, and is an equivalence of categories for $p(U) \geq 2$.

Problem 2.6. If cd (U) = r, does $H^i(U^h,G) = 0$ for every coherent analytic sheaf G on U^h, and all $i > r$?

Exercise 2.7. Show that the disc $D = \{z \in \mathbb{C} \mid |z| < 1\}$ is not isomorphic to X^h for any algebraic variety X.

Exercise 2.8. (Serre [9], p. 372). Consider the locally free rank one analytic sheaf L on $\mathbb{C}^2 - \{(0,0)\}$ defined as the structure sheaf on $U = \{z \neq 0\}$ and $V = \{w \neq 0\}$, and glued by the transition function exp (1/zw) on $U \cap V$. Show that L does not extend to a coherent sheaf on all of $\mathbb{C}^2$, and hence cannot be algebraizable.

Exercise 2.9. Let X be a complete non-singular variety over $\mathbb{C}$, and let Y be a closed subset. Let $\hat{X}$ be the formal completion of X along Y. Let X^h, Y^h be the associated analytic spaces, and let $\hat{X}^h$ be the formal completion of X^h along Y^h: as a topological space it is Y^h; its structure sheaf is $\varprojlim \mathcal{O}^h_{Y_r}$. Show that the natural map of the field of formal-rational functions $K(\hat{X})$ to the field of formal-meromorphic functions $K(\hat{X}^h)$ is an isomorphism.

Exercise 2.10. With the same notations as in the previous exercise, assume furthermore that Y is G2 in X. Then show that there is an open neighborhood N of Y^h (for the complex topology) such that the field of meromorphic functions $K(N)$ is isomorphic to $K(\hat{X}^h)$. Thus in this case, we have Y is G3 in $X \Longleftrightarrow$ every meromorphic function on a neighborhood of Y^h extends to all of X^h.

§3. Affine schemes and Stein spaces.

In this section we will show that if U is an affine scheme over $\mathbb{C}$, its associated analytic space U^h is a Stein space. Then we will give an example of Serre to show that the converse is false.

Proposition 3.1. Let U be an affine scheme of finite type over $\mathbb{C}$. Then the associated analytic space U^h is a Stein space.

Proof. We can embed U as a closed subscheme of some affine space $\mathbb{A}^n_{\mathbb{C}}$. Then U^h is a closed analytic subspace of $\mathbb{C}^n$, and hence Stein.

Example 3.2 (Serre). A non-singular algebraic surface U over $\mathbb{C}$ which is not affine, but such that its associated analytic space U^h is Stein.

Let C be a non-singular complete elliptic curve over $\mathbb{C}$. Let E be a vector bundle of rank 2 on C, which is a non-trivial extension of $\mathcal{O}_C$ by itself:

$$0 \longrightarrow \mathcal{O}_C \longrightarrow E \longrightarrow \mathcal{O}_C \longrightarrow 0 .$$

Let $X = \mathbb{P}(E)$ be the associated projective bundle. Then there is a natural projection $\pi: X \longrightarrow C$, making X into a ruled surface over C. In the following discussion, we will use the notation and results of §10, Chapter I.

The line bundle $\mathcal{O}_X(1)$ defines an effective divisor Y on X, namely the one given by the subline bundle $\mathcal{O}_C \subseteq E$. Y is the unique section of π with $(Y^2) = 0$. Let $U = X-Y$. This U is the example.

First we show that U contains no complete curves. Indeed, let Z be any irreducible curve on X. If Z is a fibre of π, $(Y.Z) = 1$. Otherwise, let us assume that Z has degree m over C, and $Z \neq mY$. Then by Proposition 10.2, Chapter I, Z corresponds to a subline bundle $M \subseteq S^m(E)$. Now E is the bundle called F_2 in the classification of Atiyah [1], Theorem 5, p. 432. He shows that on an elliptic curve in characteristic 0, $S^m(F_2) \cong F_{m+1}$ (Theorem 9, p. 438). We know also from his results that $H^0(F_m) \cong \mathbb{C}$. Thus we have either $M \cong \mathcal{O}_C$ (in which case $Z = mY$) or $\deg M < 0$. We are excluding the first case, so we conclude, again by Proposition 10.2, Chapter I, that

$$(Y.Z) = -\deg M > 0.$$

Thus U contains no complete curves.

Next we show that U is not affine. Indeed, by the theorem of Goodman (Theorem 4.2, Chapter II), it cannot be affine, because $(Y^2) = 0$, so Y cannot support any ample divisor.

Another simpler way to see that U is not affine is to calculate $H^0(U,\mathcal{O}_U)$. In fact, $H^0(U,\mathcal{O}_U) = \mathbb{C}$. There cannot be any non-constant regular function on U, because if there were, its divisor of zeroes, say Z, would be an effective curve, linearly equivalent to mY for some m, and so $(Y.Z) = 0$ which is impossible.

Now we will show that U^h is a Stein space. In fact it is analytically isomorphic to $\mathbb{C}^* \times \mathbb{C}^*$, where $\mathbb{C}^*$ denotes $\mathbb{C} - \{0\}$.

Think of C^h as a complex torus. Its universal covering space is just the complex numbers $\mathbb{C}$. We make the base extension $\mathbb{C} \longrightarrow C^h$. Since $\mathbb{C}$ is Stein, the bundle E pulled back to $\mathbb{C}$ splits, so the pull-back of the fibre bundle U^h becomes trivial, isomorphic to $\mathbb{C} \times \mathbb{C}$. The group of covering transformations is $\mathbb{Z} \oplus \mathbb{Z}$. Thus U^h is a quotient of $\mathbb{C} \times \mathbb{C}$ by a suitable action of $\mathbb{Z} \oplus \mathbb{Z}$. The orbit of $(0,0)$ cannot be contained in a single complex line, because then U^h would contain a compact curve. Such a curve would be a compact analytic subvariety of the projective analytic space X^h, hence algebraic by Chow's theorem (quoted in §2 above). But we have shown that U contains no complete curves.

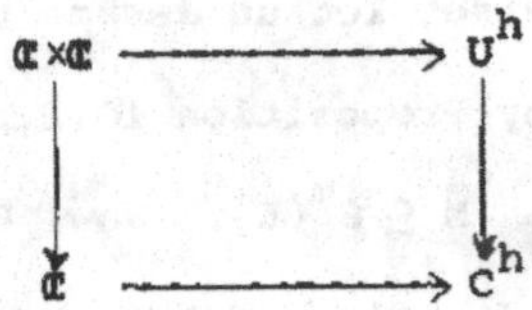

$$\begin{array}{ccc} \mathbb{C}\times\mathbb{C} & \longrightarrow & U^h \\ \downarrow & & \downarrow \\ \mathbb{C} & \longrightarrow & C^h \end{array}$$

Therefore the action of $\mathbb{Z} \oplus \mathbb{Z}$ in $\mathbb{C} \times \mathbb{C}$ is split, and the quotient is analytically isomorphic to $\mathbb{C}^* \times \mathbb{C}^*$, via the exponential map. Thus U^h is Stein.

<u>Remark</u>. This construction also furnishes an example of two algebraic varieties (namely U and $\mathbb{A}^2_{\mathbb{C}} - \{xy = 0\}$) which are not isomorphic, but whose associated analytic spaces are analytically isomorphic.

It also gives an example of a coherent sheaf (namely $\mathcal{O}_U$) such that a certain cohomology group (namely $H^o(U,\mathcal{O}_U)$) is finite-dimensional, but where the natural map

$$H^o(U,\mathcal{O}_U) \longrightarrow H^o(U^h,\mathcal{O}_{U^h})$$

is <u>not</u> an isomorphism.

Exercise 3.3. In the above example, show that if D is a divisor on X with $(D.Y) > 0$, then $H^{0}(\hat{X},\mathcal{O}_{\hat{X}}(D))$ is infinite-dimensional, where $\hat{X}$ is the completion along Y. Show also if $(D.Y) < 0$, then $H^{1}(\hat{X},\mathcal{O}_{\hat{X}}(D))$ is infinite-dimensional. Conclude that $p(U) = 0$, $q(U) = \text{cd}\,(U) = 1$.

Problem 3.4. Let X be a complete non-singular algebraic surface over $\mathbb{C}$, and let Y be an irreducible curve on X. Assume that $U = X-Y$ contains no complete curves. Give necessary and sufficient conditions on Y for U^{h} to be Stein. Is $(Y^{2}) \geq 0$ sufficient?

Exercise 3.5. Show that if one constructs the ruled surface of the above example over a field of characteristic $p > 0$, then the open set U does contain some complete curves.

§4. Negative subvarieties and the contraction problem.

Unfortunately there wasn't time to write this section. So let the reader consider the problem of giving criteria for contractibility of a subvariety. In particular, the negativity of the normal bundle is an important property. The theory presents certain interesting analogies with the theory of ample subvarieties.

For curves on surfaces, see Artin [1] and Mumford [1]. For the analytic case, see Grauert [2]. For contraction in the category of algebraic spaces, see Artin [5] and Hironaka [4]. For other related results, see Griffiths [2], Chapter V, §2 and Gieseker [1].

Index of Definitions

BIBLIOGRAPHY

Abhyankar, S.S.

[1] Tame coverings and fundamental groups of algebraic varieties - I, Amer. J. Math., 81 (1959), 46-94.

Andreotti, A.

[1] Théorèmes de dépendence algébrique sur les espaces pseudoconcaves, Bull. Soc. Math. France, 91 (1963), 1-38.

Andreotti, A. and Frankel, T.

[1] The Lefschetz theorem on hyperplane sections, Ann. of Math., 69 (1959), 713-717.

Andreotti, A. and Grauert, H.

[1] Théorèmes de finitude pour la cohomologie des espaces complexes, Bull. Soc. Math. France, 90 (1962), 193-259.

Andreotti, A. and Norguet, F.

[1] Problème de Levi pour les classes de cohomologie, C.R. Acad. Sci. Paris, 258 (1964), 778-781.

[2] Problème de Levi et convexité holomorphe pour les classes de cohomologie, Ann. Sc. Norm. Sup. Pisa, 20 (1966), 197-241.

[3] La convexité holomorphe dans l'espace analytique des cycles d'une variété algébrique, Ann. Sc. Norm. Sup. Pisa, 21 (1967), 31-82.

Andreotti, A and Salmon, P.

[1] Anelli con unica decomponibilità in fattori primi ed un problema di intersezioni complete, Monatshefte für Math., 61 (1957), 97-142.

Artin, M.

[1] Some numerical criteria for contractibility of curves on algebraic surfaces, Amer. J. Math., 84 (1962), 485-496.

[2] The implicit function theorem in Algebraic Geometry, Proc. of the Bombay Colloquium on Algebraic Geometry, (1968), 13-34.

[3] Algebraic approximation of structures over complete local rings, Publ. Math. IHES, 36 (1969), 23-58.

[4] Algebraization of formal moduli - I, A Collection of Mathematical papers in Honour of K. Kodaira, University of Tokyo Press, (to appear).

[5] Algebraization of formal moduli - II (Existence of modifications), Ann. of Math., 91 (1970), 88-135.

Artin, M. and Grothendieck, A.

[SGAA] Cohomologie étale des schémas, mimeographed notes, IHES, Paris, (1963-64).

Atiyah, M.F.

[1] Vector bundles over an elliptic curve, Proc. Lond. Math. Soc. (3), 7 (1957), 414-452.

Atiyah, M.F. and Hodge, W.V.D.

[1] Integrals of the second kind on an algebraic variety, Ann. of Math., 62 (1955), 56-91.

Barth, W.

[1] Fortsetzung meromorpher Funktionen in Tori und Komplex-projektiven Räumen, Invent. Math., 5 (1968), 42-62.

[2] Der Abstand von einer algebraischen Mannigfaltigkeit in Komplex-projektiven Raum (to appear).

[3] Uber die analytische Cohomologiegruppe $H^{n-1}(\mathbb{P}^n - A, F)$, Invent. Math., 9 (1970), 135-144.

[4] Transplanting cohomology classes in complex-projective space (to appear).

Barton, C.M.

[1] Tensor products of ample vector bundles in characteristic p (to appear).

Borel, A. and Serre, J.-P.

[1] Le Théorème de Riemann-Roch, Bull. Soc. Math. France, 86 (1958), 97-136.

Borelli, M.

[1] Divisorial Varieties, Pacific J. Math., 13 (1963), 375-388.

[2] Some results on ampleness and divisorial schemes, Pacific J. Math., 23 (1967), 217-227.

[3] Affine Complements of Divisors, Pacific J. Math., 31 (1969), 595-607.

Bott, R.

[1] On a theorem of Lefschetz, Mich. Math. J., 6 (1959), 211-216.

Cartan, H.

[1] Variétés analytiques complexes et cohomologie, Colloque sur les fonctions de plusieurs variables, C.B.R.M. Bruxelles, (1953), 41-55.

Chevalley, C.

[1] Sur la théorie de la variété de Picard, Amer. J. Math., 82 (1960), 435-490.

Chow, W.L.

[1] On compact complex analytic varieties, Amer. J. Math., 71 (1949), 893-914.

[2] The fundamental group of an algebraic variety, Amer. J. Math., 74 (1952), 726-736.

[3] On meromorphic maps of algebraic varieties, Ann. of Math., 89 (1969), 391-403.

Deligne, P. and Mumford, D.

[1] The irreducibility of the space of curves of given genus, Publ. Math. IHES, 36 (1969), 75-109.

Dieudonné, J.

[1] Lie groups and Lie hyperalgebras over a field of characteristic $p > 0$ - II, Amer. J. Math., 76 (1955), 218-244.

Dolbeault, P.

[1] Sur la cohomologie de variétés analytiques complexes, C.R. Acad. Sci. Paris, 236 (1953), 175-177.

Fano, G.

[1] Sulle varietà algebriche che sono intersezioni complete di più forme, Atti R. Acc. Torino, 44 (1909), 633-648.

Fáry, I.

[1] Cohomologie des variétés algébriques, Ann. of Math., 65 (1957), 21-73.

Forster, O. and Ramspott, K.J.

[1] Analytische Modulgarben und Endromisbündel, Invent. Math., 2 (1966), 145-170.

Franchetta, A.

[1] Sulle curve appartenenti a una superficie generale d'ordine $n \geq 4$ dell S_3, Atti Acc. dei Lincei, Rend. (8), 3 (1947), 71-78.

Frisch, J. and Guenot, J.

[1] Prolongement de faisceaux analytiques cohérents, Invent. Math., 7 (1969), 321-343.

Fulton, W.

[1] Algebraic Curves, Lecture Notes in Mathematics (Notes by Richard Weiss), Benjamin, (1969).

Gieseker, D.

[1] Contributions to the theory of positive embeddings in algebraic geometry, Thesis, Harvard University, Cambridge, (1969).

Godement, R.

[1] Topologie Algébrique et Théorie des Faisceaux, Hermann Paris, (1958).

Goodman, J.E.

[1] Affine open subsets of algebraic varieties and ample divisors, Ann. of Math., 89 (1969), 160-183.

Goodman, J.E. and Hartshorne, R.

[1] Schemes with finite-dimensional cohomology groups, Amer. J. Math., 91 (1969), 258-266.

Grauert, H.

[1] Une notion de dimension cohomologique dans la théorie des espaces complexes, Bull. Soc. Math. France, 87 (1959), 341-350.

[2] Uber Modifikationen und exzeptionelle analytische Mengen, Math. Ann., 146 (1962), 331-368.

Griffiths, P.A.

[1] Hermitian differential geometry and the theory of positive and ample holomorphic vector bundles, J. Math. and Mech., 14 (1965), 117-140.

[2] The extension problem in complex analysis - II, (Embeddings with positive normal bundle), Amer. J. Math., 88 (1966), 366-446.

Grothendieck, A.

[1] Sur quelques points d'algèbre homologique, Tôhoku Math. J., 9 (1957), 119-221.

[2] Sur les faisceaux algébriques et les faisceaux analytiques cohérents, Exposé 2, Séminaire H. Cartan, 9 (1956/57).

[3] Fondements de la géométrie algébrique, extraits du sem. Bourbaki (1957-62), mimeo. notes, Secr. Math., Paris, (1962).

[SGA 1] Séminaire de Géométrie Algébrique; Revêtements étales et groupe fondemental, mimeographed notes, IHES, Paris, (1960-61).

[SGA 2] Cohomologie locale des faisceaux cohérents et théorèmes de Lefschetz locaux et globaux (SGA 2), North-Holland Publishing Company - Amsterdam, (1968).

[LC] Local Cohomology, Lecture notes in Mathematics 41, Springer (1966). (Notes by R. Hartshorne).

[7] On the De Rham cohomology of algebraic varieties, Publ. Math. IHES, 29 (1966), 95-103.

Grothendieck, A. and Dieudonné, J.

[EGA] Éléments de géométrie algébrique, Publ. Math. IHES, 4, 8, 11, 17, 20, 24, 28, and 32 (1960-1967).

Gunning, R.C. and Rossi, H.

[1] Analytic functions of several complex variables, Prenctice-Hall, Englewood Cliffs, N.J., (1965).

Hartshorne, R.

[1] Complete intersections and connectedness, Amer. J. Math., 84 (1962), 497-508.

[AVB] Ample vector Bundles, Publ. Math. IHES, 29 (1966), 63-94.

[RD] Residues and Duality, Lecture Notes in Mathematics 20, Springer (1966).

[CDAV] Cohomological Dimension of Algebraic Varieties, Ann. of Math., 88 (1968), 403-450.

[5] Curves with high self-intersection on algebraic surfaces, Publ. Math. IHES, 36 (1969), 111-125.

[6] Affine duality and cofiniteness, Invent. Math., 9 (1970), 145-164.

[7] Ample vector bundles on curves (to appear).

Hironaka, H.

[1] On the theory of birational blowing-up, Thesis (unpublished), Harvard University, Cambridge, (1960).

[2] An example of a non-kählerian complex-analytic deformation of kählerian complex structures, Ann. of Math., 75 (1962), 190-208.

[3] Resolution of singularities of an algebraic variety over a field of characteristic zero I-II, Ann. of Math., 79 (1964), 109-326.

[4] Formal line bundles along exceptional loci, Proc. of the Bombay Colloquium on Algebraic Geometry, (1968), 201-218.

[5] Smoothing of algebraic cycles of small dimensions, Amer. J. Math., 90 (1968), 1-54.

[6] On some formal imbeddings, Ill. J. Math., 12 (1968), 587-602.

Hironaka, H. and Matsumura H.

[1] Formal functions and formal embeddings, J. Math. Soc. Japan, 20 (1968), 52-82.

Hirzebruch, F.

[1] Der Satz von Riemann-Roch in Faisceau-theoretischer Formulierung, einige Anwendungen und offene Fragen, Proc. Int. Cong. Math., Amsterdam, Vol. 3, (1954), 457-473.

Hodge, W.V.D.

[1] The topological invariants of algebraic varieties, Proc. Int. Cong. Math. Cambridge, (1950), 182-192.

Igusa, J.-I.

[1] On some problems in abstract algebraic geometry, Proc. Nat. Acad. Sci., USA, 41 (1955), 964-967.

Kaljulaid, U.

[1] On the cohomological dimension of some quasi-projective varieties (in Russian), Izv. Akad. Nauk Est. SSR, 18 (1969), 261-272.

Kleiman, S.L.

[1] A Note on the Nakai-Moisezon test for ampleness of a divisor, Amer. J. Math., 87 (1965), 221-226.

[2] Towards a numerical theory of ampleness, Ann. of Math., 84 (1966), 293-344.

[3] On the vanishing of $H^n(X,F)$ for an n-dimensional variety, Proc. Amer. Math. Soc., 18 (1967), 940-944.

[4] Geometry on Grassmannians and applications to splitting bundles and smoothing cycles, Publ. Math. IHES, 36 (1969), 281-297.

[5] Ample vector bundles on surfaces (to appear).

Kneser, M.

[1] Uber die Darstellung algebraischer Raumkurven als Durchschnitte von Flächen, Archiv der Math., 11 (1960), 157-158.

Knutson, D.

[1] Algebraic Spaces, Thesis, MIT, (1968), (to appear).

Kodaira, K.

[1] On a differential-geometric method in the theory of analytic stacks, Proc. Nat. Acad. Sci. USA, 39 (1953), 1268-1273.

Kodaira, K. and Spencer, D.

[1] On a theorem of Lefschetz and the Lemma of Enriques-Severi-Zariski, Proc. Nat. Acad. Sci. USA, 39 (1953), 1273-1278.

Lang, S. and Néron, A.

[1] Rational points of abelian varieties over function fields, Amer. J. Math., 81 (1959), 95-118.

Laufer, H.B.

[1] On Serre duality and envelopes of holomorphy, Trans. Amer. Math. Soc., 128 (1967), 414-436.

Lefschetz, S.

[1] On certain numerical invariants of algebraic varieties, Trans. Amer. Math. Soc., 22 (1921), 326-363.

[2] L'analysis situs et la géométrie algébrique, Gauthier-Villars, Paris, (1924).

Leray, J.

[1] Résidus, Sem. Bourbaki, no. 183, Paris, (1959).

Lipman, J.

[1] Rational singularities, with applications to algebraic surfaces and unique factorization, Publ. Math. IHES, 36 (1969), 195-279.

Macdonald, I.G.

[1] Algebraic Geometry (Introduction to Schemes), Benjamin, (1968).

Moisezon, B.G.

[1] A criterion for projectivity of complete algebraic varieties, AMS translations (2), 63 (1967), 1-50.

[2] On n-dimensional compact varieties with n algebraically independent meromorphic functions, AMS translations (2), 63 (1967), 51-177.

[3] On algebraic cohomology classes on algebraic varieties, (in Russian), Izvestia Akad. Nauk USSR, Ser. Math., 31 (1967), 225-268.

Monsky, P.

[1] Formal cohomology II, Ann. of Math., 88 (1968), 218-238.

[2] Finiteness of De Rham cohomology (to appear).

Mumford, D.

[1] The topology of normal singularities of an algebraic surface and a criterion for simplicity, Publ. Math. IHES, 9 (1961), 5-22.

[2] Pathologies of modular algebraic surfaces, Amer. J. Math., 83 (1961), 339-342.

[3] Lectures on curves on an algebraic surface, Ann. of Math. Studies 59, Princeton, (1966).

[4] Pathologies III, Amer. J. Math., 89 (1967), 94-104.

Murre, J.P.

[1] Lectures on "An Introduction to Grothendieck's Theory of the Fundamental Group", Tata Institute Lecture Notes, Bombay, (1967).

Nagata, M.

[1] Remarks on a paper of Zariski on the purity of branch-loci, Proc. Nat. Acad. Sci. USA, 44 (1958), 796-799.

[2] Existence theorems for non-projective complete algebraic varieties, Ill. J. Math., 2 (1958), 490-498.

[3] On rational surfaces I, Mem. Coll. Sci., Kyoto, A 32 (1960), 351-370.

[4] Imbedding of an abstract variety in a complete variety, J. Math. Kyoto Univ., (1962), 1-10.

[5] Local Rings, Interscience tracts 13, Wiley, New York, (1962).

Nakai, Y.

[1] On the characteristic linear systems of algebraic families, Ill. J. Math., 1 (1957), 552-561.

[2] A criterion of an ample sheaf on a projective scheme, Amer. J. Math., 85 (1963), 14-26.

[3] Some fundamental lemmas on projective schemes, Trans. Amer. Math. Soc., 109 (1963), 296-302.

Narasimhan, M.S. and Seshadri, C.S.

[1] Stable and unitary vector bundles on a compact Riemann surface, Ann. of Math., 82 (1965), 540-567.

Narasimhan, R.

[1] The Levi problem for complex spaces, Math. Ann., 142 (1961), 355-365.

[2] The Levi problem for complex spaces II, Math. Ann., 146 (1962), 195-216.

Noether, M.

[1] Zur Grundlegung der Theorie der algebraischen Raumcurven, Berliner Abh., (1882).

Picard, E. and Simart,

[1] Théorie des fonctions algébriques de deux variables indépendentes, Paris, (1897).

Rossi, H.

[1] Continuation of subvarieties of projective varieties, Amer. J. Math., 91 (1969), 565-575.

Samuel, P.

[1] Méthodes d'algèbre abstraite en géométrie algébrique, Ergebn. der Math., Springer, (1955).

Scheja, G.

[1] Fortsetzungssätze der Komplex-analytischen Cohomologie und ihre algebraische Charakterisierung, Math. Ann., 157 (1964), 75-94.

Schwarzenberger, R.L.E.

[1] Vector bundles on algebraic surfaces, Proc. Lond. Math. Soc. (3), 11 (1961), 601-622.

Serre, J.-P.

[1] Quelques problèmes globaux relatifs aux variétés de Stein, Colloque sur les fonctions de plusieurs variables, C.B.R.M. Bruxelles, (1953), 57-68.

[2] Un théorème de dualité, Comment. Math. Helv., 29 (1955), 9-26.

[FAC] Faisceaux Algébriques Cohérents, Ann. of Math., 61 (1955), 197-278,

[GAGA] Géométrie Algébrique et Géométrie Analytique , Ann. Inst. Fourier, 6 (1956), 1-42.

[5] Sur la cohomologie des variétés algébriques, J. Math. Pures et Appl., 36 (1957), 1-16.

[6] Sur la topologie des variétés algébriques en caractéristique p, Symposium International de Topologia Algebraica, Mexico, (1958), 24-53.

[7] Morphismes universels et variétés d'Albanese, Exposé 10, Séminaire C. Chevalley: Variétés de Picard, 3 (1958/59).

[8] Groupes Algébriques et Corps de Classes, Hermann, Paris, (1959).

[9] Algèbre Locale-Multiplicités, Lecture Notes in Mathematics 11, Springer, (1965).

[10] Prolongement de faisceaux analytiques cohérents, Ann. Inst. Fourier, 16 (1966), 363-374.

Seshadri, C.S.

[1] L'opération de Cartier. Applications. Exposé 6 Séminaire C. Chevalley: Variétés de Picard, 3 (1958/59).

Severi, F.

[1] Una proprietà delle forme algebriche prive di punti multipli, Rend. dei Lincei (5), 15 (1906), 691-696.

Siu, Y.-T.

[1] Extending coherent analytic sheaves, Ann. of Math., 90 (1969), 108-143.

[2] Analytic sheaves of local cohomology, Bull. Amer. Math. Soc., 75 (1969), 1011-1012.

Snapper, E.

[1] Multiples of Divisors, J. Math. and Mech., 8 (1959), 967-992.

[2] Polynomials Associated with Divisors, J. Math. and Mech., 9 (1960), 123-129.

Sorani, G. and Villani, V.

[1] q-complete spaces and cohomology, Trans. Amer. Math. Soc., 125 (1966), 432-448.

Spanier, E.H.

[1] Algebraic Topology, McGraw-Hill, N.Y., (1966).

Speiser, R.D.

[1] Cohomological Dimension of Abelian Varieties, Thesis, Cornell University, (1970).

Suominen, K.

[1] Duality for coherent sheaves on analytic manifolds, Ann. Acad. Sci. Fenn. Ser (A), 424 (1968), 1-19.

Trautmann, G.

[1] Cohérence de faisceaux analytiques de la cohomologie locale, C.R. Acad. Sci. Paris, 267 (1968), 694-695.

[2] Ein Endlichkeitsatz in der analytischen Geometrie, Invent. Math., 8 (1969), 143-174.

Vesentini, E.

[1] Lectures on Levi convexity of complex manifolds and cohomology vanishing theorems, Lecture Notes, Tata Institute of Fundamental Research, Bombay, (1967).

Villani, V.

[1] Cohomological properties of complex spaces which carry over to normalizations, Amer. J. Math., 88 (1966), 636-645.

Wallace, A.H.

[1] Homology theory of algebraic varieties, Pergamon, (1958).

Zariski, O.

[1] A theorem on the Poincaré group of an algebraic hypersurface, Ann. of Math., 38 (1937), 131-141.

[2] Theory and applications of holomorphic functions on algebraic varieties over arbitrary ground fields, Memoirs, AMS, (1951).

[3] Complete linear systems on normal varieties and a generalization of a lemma of Enriques-Severi, Ann. of Math., 55 (1952), 552-592.

[4] Introduction to the problem of minimal models in the theory of algebraic surfaces, Publ. Math. Soc. Japan, 4 (1958).

[5] On the purity of the branch locus of algebraic functions, Proc. Nat. Acad. Sci. U.S.A., 44 (1958), 791-796.

[6] The theorem of Riemann-Roch for high multiples of an effective divisor on an algebraic surface, Ann. of Math., 76 (1962), 560-615.

Zariski, O. and Samuel, P.

[1] Commutative Algebra, Vol. I and II, Van Nostrand, Princeton, (1958, 1960).